Plants of the Adelaide Plains and Hills

Plants of the Adelaide Plains and Hills

Gilbert R.M. Dashorst
and
John P. Jessop

State Herbarium of South Australia

Kangaroo Press

Dashorst, Gilbert R.M. (Gilbert Roelof Maria), 1956- .
Plants of the Adelaide plains and hills.

Bibliography.
Includes index.
ISBN 0 86417 323 7.

1. Botany - South Australia - Adelaide Region. 2. Botany - South Australia - Adelaide Region - Pictorial works. I. Jessop, J. P. (John Peter), 1939- . II. Title.

581.994232

First published in 1990 by Kangaroo Press Pty Ltd
3 Whitehall Road (P. O. Box 75) Kenthurst NSW 2156
Typeset by G. T. Setters Pty Limited
Printed in Singapore by Kyodo Printing Co (S'pore) Pte Ltd

Contents

Foreword

In March 1988 Gilbert Dashorst came to Europe for five months under the sponsorship of the 'Winston Churchill Memorial Trust' when he showed me some of his early work.

After four weeks in the United Kingdom, where he took great advantage of the libraries opened to him through the 'Trust', he went to study practical botanical illustration at the Leiden Herbarium.

Gilbert Dashorst had followed in the footsteps of Georg Dionysius Ehret who went to study there in 1736 before settling in London in 1737. No finer shade could inspire a modern botanical artist.

Less than one year after meeting Gilbert Dashorst I met him again at the Adelaide Botanic Gardens Herbarium and saw the work he had completed.

It is much in the artist's interest to work with professional botanists and seek their advice and help, as this book clearly demonstrates.

I am delighted to be asked to write the Foreword to a book with such lively, well-drawn and beautifully-designed pages, completed from living material over three years.

Margaret Stones

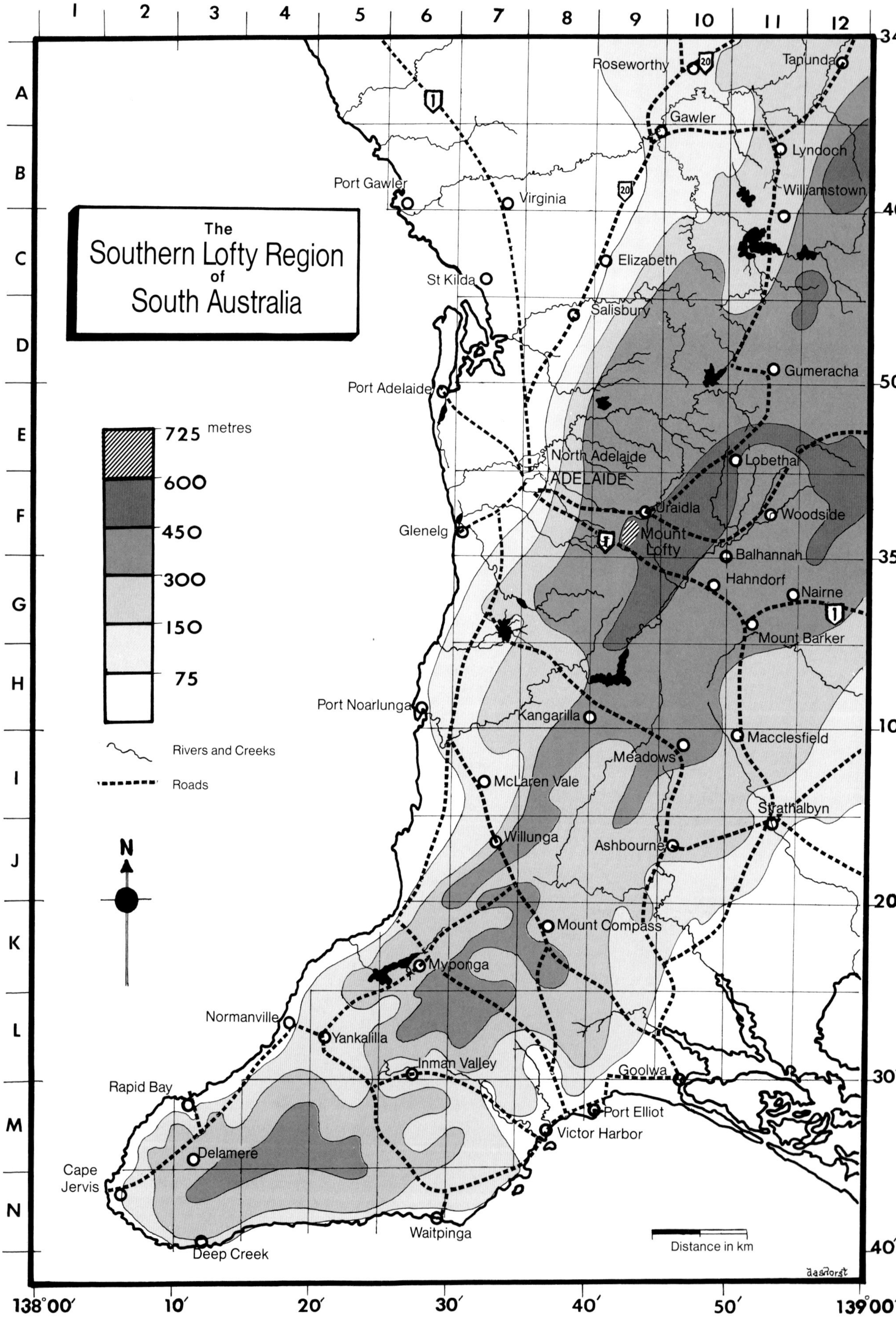
The
Southern Lofty Region
of
South Australia
725 metres
600
450
300
150
75
Rivers and Creeks
Roads
N
Distance in km
Roseworthy
Tanunda
Gawler
Lyndoch
Williamstown
Port Gawler
Virginia
Elizabeth
St Kilda
Salisbury
Gumeracha
Port Adelaide
North Adelaide
ADELAIDE
Lobethal
Uraidla
Woodside
Glenelg
Mount Lofty
Balhannah
Hahndorf
Nairne
Mount Barker
Port Noarlunga
Kangarilla
Macclesfield
Meadows
McLaren Vale
Strathalbyn
Willunga
Ashbourne
Mount Compass
Myponga
Normanville
Yankalilla
Inman Valley
Goolwa
Rapid Bay
Port Elliot
Victor Harbor
Delamere
Cape Jervis
Waitpinga
Deep Creek
1 2 3 4 5 6 7 8 9 10 11 12
A B C D E F G H I J K L M N
34°30′
40′
50′
35°00′
10′
20′
30′
40′
138°00′
10′
20′
30′
40′
50′
139°00′
daaHorst

Introduction

Plants of the Adelaide Plains and Hills was planned to help people to appreciate and learn more about the plants of the Adelaide area.

It is very much hoped that it will contribute to the pleasure people get from these plants and that it will be a reminder for them of the flora being lost to housing, farming, industry and pollution. Even in selecting plants previously quite commonly collected here, it was realised that many are now much less common or perhaps extinct. Similarly, introduced 'weedy' species unfortunately form an ever more important proportion of our common plants. We are warned that climatic changes may accelerate these processes. So enjoy the flora while you can and do what you can to enable others, who come after you, to do so too.

Scope

About two thousand flowering plant and fern species grow wild in the area covered by this book. I have aimed at illustrating about half of these: the one thousand most frequently collected. In addition, a small selection of lower plants, mushrooms, algae (seaweeds), lichens, liverworts and mosses has been included.

The style of illustration I have used is rare in works of this kind, but I believe that the combination of pencil and colour has aesthetic merit. More importantly, however, it has enabled me to draw representative parts of the plant while emphasising particular features by the use of colour. Any detail shown on the plates should be visible to the naked eye or with the aid of a 10-power hand lens.

The area covered by this work (map p.11) is called the Southern Lofty Region in *Flora of South Australia*. Although not defined by any natural features, it includes the most populated area of the State and therefore the places where people are most likely to find plants. Use of this region also had practical advantages in terms of the availability of species lists for planing the book.

Identification

How does one make it as easy as possible for a beginner to identify a plant they have found? There are many answers, although no one by itself provides a satisfactory solution. Compared with, say, birds, plants are often not easy to identify, but illustrations in colour are surely the single most helpful thing a book can provide. Should the species be arranged according to colour, which is not consistent for many species, not uniform even on a single flower and not always recognised by everyone as the same colour? What about times of the year (although flowering times vary according to weather)? One could try habitat, or size, or shape, or structure of flowers. No single factor is easy to observe (without picking the flower, preferably) and at the same time consistent.

In the end it was decided to arrange most of the plates by plant families. Several families will be quite easy to recognise after looking at the plates a few times, e.g., Orchidaceae (orchids), Compositae (daisies) and Proteaceae (hakeas and banksias). Acquiring a feeling for their classification is the best start to identifying plants one has not seen before. If, as I hope, people using this book become sufficiently interested and confident, this basic knowledge of families will be invaluable when they turn to the *Flora of South Australia*, which follows the same sequence, for more information.

Organisations such as the WEA have from time to time provided introductory courses in plant identification. Attendance at such a course should be very helpful in introducing the beginner to the structure of wildflowers and to their classification.

Text

For the lower plants the family name is given after the plant name except for the algae where the order name is given. For the ferns and flowering plants the family names provide a heading to a number of species.

There is little text; the plates provide most of what is needed. The scientific name is preceeded by an asterisk if the species has been introduced to the area. It is followed by one common name where known, although others may exist. The names used here are as accurate as could be achieved at the time of going to press, but as research continues, changes in names will become necessary.

The descriptive text is largely to provide measurements. The sometimes enormous range in the size of a plant or of one of its organs can only be expressed in figures which is one reason why scales are not included on the plates. Why give the scale for the single leaf drawn, when, for example, leaves may vary from 5–30 mm in other specimens? In some cases further information is given to help with identification.

Distribution is given in terms of 12 other regions of South Australia (map p.11), followed by interstate (abbreviated) and overseas areas. There is a statement of the place of origin of introduced species. Distribution information was not available for the fungi.

It has sometimes been possible to give a simple way of recognising other similar and related species not illustrated. In other cases there is only a reference to the number of species of the same genus not illustrated. Note that this number refers only to other species in the area covered by the book, not to species occurring outside the area. This follows the last illustrated species of each genus, but has sometimes not been available, especially for the fungi.

Technical terms have been all but eliminated. Where they have been used, reference to the drawing should make it obvious what is meant. For example, look at plate XCIII, fig. 3b: although there is no explanation of its botanical nature it is easy to see which part of the flower is called a 'labellum'. The short glossary on p.14 includes the few abbreviations.

Distribution Maps

Small maps show the distribution of each species within our area. These were plotted over a number of years from specimens in the State Herbarium of South Australia. Because of delays in processing specimens, these maps, despite considerable efforts, do not therefore show every locality available when going to press. Nevertheless, they do reasonably reflect the known distribution of each species. The algae (seaweeds) were not plotted as they are often found as drift on the beach rather than growing attached in the water and the fungal maps are based on very limited information.

The map on p.8 indicates some landmarks of use in locating localities on the small distribution maps. Where an illustrated plant has been identified to genus but not species, the locality mapped is that of the specimen illustrated; other similar specimens already in the Herbarium collections have not been mapped.

The maps serve two purposes: firstly, they provide information on the known distribution; secondly, they afford an opportunity for the reader to contribute to knowledge of the distribution of our flora.

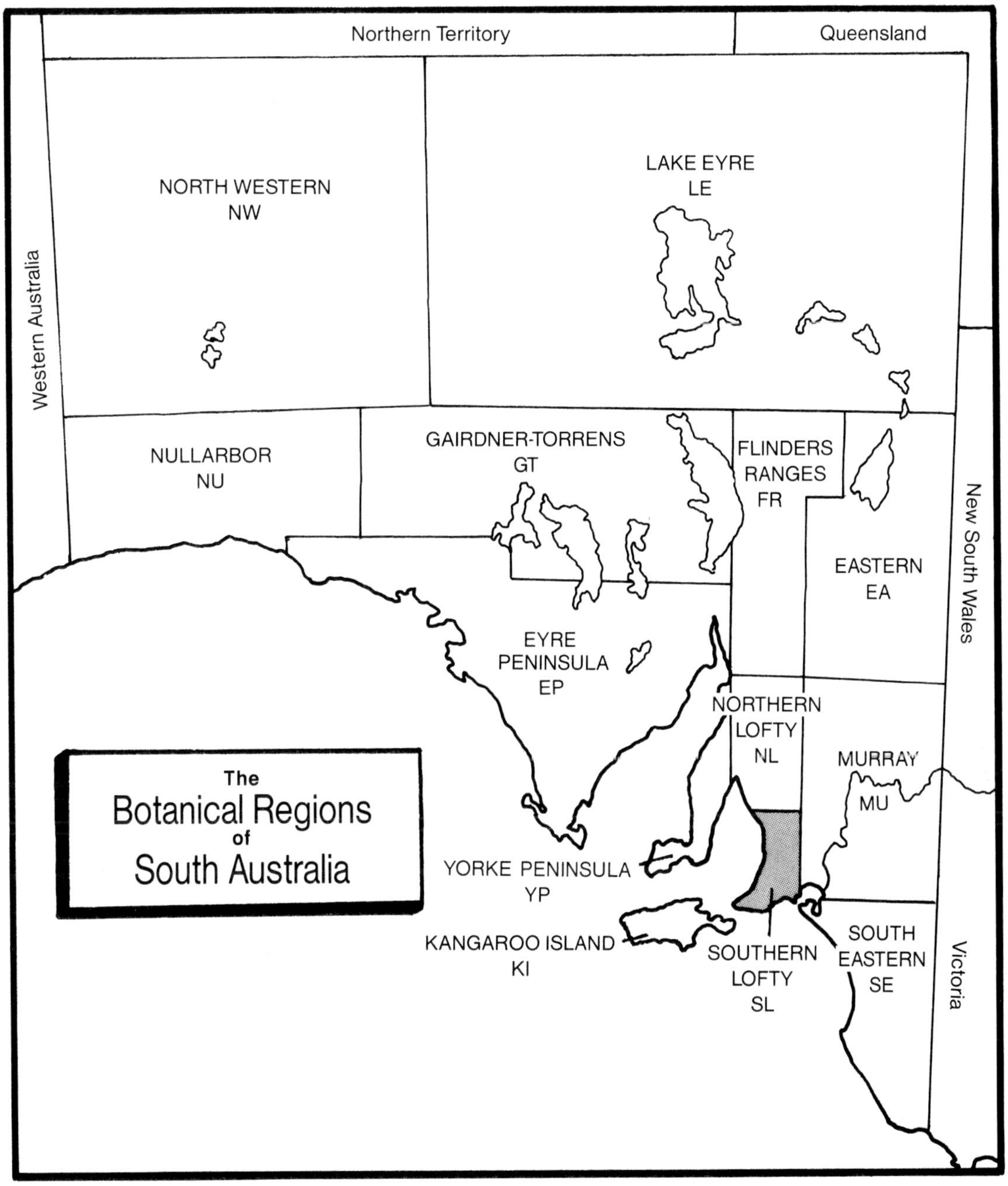

What You Can Do

1. Use this book to identify the plant you have found.

2. If you find it, make sure the locality is in one of the filled in blocks on its map. The gazetteer on page 208 will help you to locate the square in which the plant was found.

3. If it is not recorded on the map, collect it if you are legally allowed to (check with the National Parks and Wildlife Service) and if removal of your specimen cannot possibly jeopardise the survival of the species in that area. A free leaflet is available from the State Herbarium (Adelaide Botanic Garden, North Terrace) advising on what should be collected and how it should be labelled and preserved. This specimen should then be submitted to the Herbarium to add to its record of the distribution of the flora.

4. If you cannot identify the specimen, but can legally and safely collect it, you may press and label it and donate it to the Herbarium which in return should be able to identify it for you. As the common species are in the book, anything not found in it may prove to be of interest.

Edible Mushrooms

Many wild mushrooms are edible; others are poisonous. This book, with its incomplete selection of species, does not provide a sufficiently reliable identification for safe use in deciding what one can eat. This is why I have not indicated in the text which can be eaten.

Acknowledgments

Many people have been involved in this project since its inception in 1986: people helping with collecting live material for painting, small groups helping with mapping, and those who helped with plant identification and the text.

A special thank-you is due to D. Murfet for the time spent and the kilometres travelled collecting fresh plant material for illustrating. While most illustration for botanical work is in black and white and can therefore be done from preserved specimens, the use here of colour has meant fresh specimens and a great dependence on help from collectors.

Others whom I would like to thank for assisting in collecting and identifying plants as well as checking the illustrations are: C.R. Alcock, R.M. Barker, W.R. Barker, R. Bates, G.H. Bell, K.J. Bellette, C.M. Birchby, R.J. Chinnock, R. Coles, D.D. Cunningham, N.N. Donner, D. Fotheringham, D. Frappell, E.C. Foster, C.A. Grgurinovic, L. Haegi, J. Heap, D. Hopton, D. Keane, D.N. Kraehenbuehl, A. Michelmore, B. Moore, A.A. Munir, M.C. O'Leary, L. Padgham, A. Prescott, B.R. Randell, E.L. Robertson, N.M. Smith, D.E. Symon, R.L. Taplin, H.R. Toelken, H.P. Vonow, J.Z. Weber, D.J.E. Whibley, H.B.S. Womersley and J. Woodman.

Much help was given by the following people with extracting plant localities from herbarium records and mapping them. Thanks to M. van Velsen and her helpers B. Bowden, M. Bolton, P. Grulich, U. Roberts and H. Ward. Assistance was also given by the staff from the Adelaide, Mt Lofty and Wittunga Botanical Gardens in advising me on what plants were available at these sites.

The Board of the Adelaide Botanic Gardens made the time and facilities available for undertaking the project.

Final thanks go to M.T. Eadsforth for typing the manuscript, M. Stones, the Australian born botanical illustrator, for writing the Foreword and all the other friends and people who supported me by their words of encouragement and suggestions during this project.

G.R.M. Dashorst
State Herbarium of South Australia

Glossary

lf—leaf

lvs—leaves

fl—flower

fls—flowers

infl—inflorescence; the spike, head or any other arrangement of the flowers.

family—a major group of genera and species. For example, the Myrtaceae is a family including the genera *Eucalyptus* (gums) and *Leptospermum* (tea-trees) and *Callistemon* (bottle brushes). Usually ends -aceae.

genus (plural genera)—a group of species. For example, *Banksia*, which has two species, *Banksia marginata* and *Banksia odorata*.

order—a group of families (ending -ales). Although every family belongs to an order, in most plant groups orders are not often used by botanists.

species (plural unchanged)—a particular kind of plant such as 'pink gum' or 'river red gum'. Each species has a scientific two-word name; the generic name, such as *Eucalyptus*, and the specific epithet. These two gums are *Eucalyptus fasciculosa* and *Eucalyptus camaldulensis*. Note that when it is not ambiguous the generic name can be abbreviated to a single letter, so in this case the second species could be written *E. camaldulensis*.

'. . .' quoted terms—terms are placed in quotes when they are not strictly correct botanically. For example, in some of the Chenopodiaceae the term 'fruit' is used for a structure made up of the fruit and also some surrounding organs. 'Leaves' in *Acacia* are usually called phyllodes, botanically, because they have evolved from leaves by a reduction to little more than the original leaf stalk.

* **asterisked plants**—introduced to the area and gone wild.

♂ — male

♀ — female

⚥ — bisexual

Literature

CATCHESIDE, D.G. (1980). *Mosses of South Australia*. Government Printer, Adelaide.

CLELAND, J.B. (1934-35). *Toadstools and Mushrooms and Other Larger Fungi of South Australia*, vols I-II. Government Printer, Adelaide. Reprinted in 1976.

DUNCAN, B.D. & ISAAC, G. (1986). *Ferns and Allied Plants of Victoria, Tasmania and South Australia*. Melbourne University Press, Carlton.

FILSON, R.B. & ROGERS, R.W. (1979). *Lichens of South Australia*. Government Printer, Adelaide.

JESSOP, J.P. & TOELKEN, H.R. Eds (1986). *Flora of South Australia*, vols I-IV. Government Printer, Adelaide. Descriptions of the flowering plants and ferns of South Australia.

PRESCOTT, A. (1988). *It's Blue with Five Petals*. The author, Prospect. A field guide to wildflowers of the Adelaide region.

SCOTT, G.A.M. (1985). *Southern Australian Liverworts*. Australian Government, Canberra.

SCOTT, G.A.M. & STONE, I.G. (1976). *The Mosses of Southern Australia*. Academic Press, London.

WOMERSLEY, H.B.S. (1984-87). *The Marine Benthic Flora of Southern Australia*, pts I-II. Government Printer, Adelaide. Part III will conclude the coverage of the sea-grasses and those algae which are not free-floating.

YOUNG, A.M. (1986). *Common Australian Fungi*. New South Wales University Press, Kensington.

Plate I

CHLOROPHYTA (green algae)

1. *Chaetomorpha coliformis,* mermaids necklace, CLADOPHORALES: Long unbranched chains (to 40 cm long) of globular cells 1-4 mm diam., glistening-green; usually on the seagrasses *Amphibolis* or *Heterozostera*, or on brown or red tufted algae, under moderate wave action. Eyre Peninsula to Walkerville, Vic. and around Tas. (Victor Harbor region). Approx. 4 other species.

2. *Codium pomoides,* CODIALES: Firm dark green globular balls 4–12 cm diam., with a smooth surface, tightly adherent to rock from low-tide level (shaded) to 20 m deep, on rough-water coasts. Recherche Archipelago, WA, to Walkerville, Vic., and around Tas. (Fleurieu Peninsula). Approx. 7 other species.

3. *Codium mamillosum,* CODIALES: Light green moderately firm (when living) globular balls 2–8 cm diam., with slightly projecting ends of filaments (utricles) on the surface; on rock from low-tide level (shaded) to 40 m deep, under moderate water movement. Dongara, WA, to Warrnambool, Vic. (Fleurieu Peninsula to Adelaide beaches). Approx. 7 other species. (See also Fig. 11.)

4. *Enteromorpha* sp., ULVALES: Light green hollow tubes 2–10 cm high, expanding upwards, delicate, often slightly convolute; on rock or floating structures, near low-tide level, appearing and growing quickly in response to nutrients (from land run-off) and dying away after a few weeks. Cosmopolitan; on all southern Australian coasts. Approx. 5 other species.

5. *Ulva* sp., ULVALES: Light to dark green (sometimes yellowish) flat or convolute sheets usually 4–40 cm high and across, thin and delicate; on rocks or solid substrates at and just below low-tide level, responding with increased growth and much larger size to nutrients (e.g., at St Kilda), and dying after a few weeks. Cosmopolitan; on all southern Australian coasts. Approx. 5 other species.

6. *Caulerpa trifaria,* CAULERPALES: Light to medium green, with erect fronds 5–25 cm high, with 3 rows of short cylindrical laterals 3–10 mm long, arising from a basal stolon; attached to rock or firm substrate in moderately sheltered water 2–31 m deep. Southern Australia generally (Gulf St Vincent). Approx. 10 other species.

7. *Caulerpa cactoides,* CAULERPALES: Medium to dark green, with erect fronds 10–25 cm high each with 2 rows of opposite club-shaped laterals (slimy when broken) 1–2 cm long, arising from a basal stolon with rhizoids; on rock or in sandy-mud, from rough-water to calm, 0–38 m deep. Southern Australia generally (Victor Harbor to Adelaide beaches). Approx. 10 other species.

8. *Dictyosphaeria sericea,* SIPHONOCLADALES: Medium green, shallowly cup-shaped to lobed, 2–10 cm across, relatively thin; attached to rock by the central basal region where large cells occur. Southern Australia generally (Victor Harbor to Marino).

9. *Bryopsis plumosa,* DERBESIALES: Light green, delicate, much branched, 2–15 cm high, growing on rock or solid substrates under calm to moderate water movement, from low-tide level to several metres deep. Cosmopolitan; on most southern Australian coasts (Gulf St Vincent). Approx. 2 other species.

10. *Caulerpa scalpelliformis,* CAULERPALES: Medium green with erect fronds 4–10 cm high, bearing two rows of scalpel-shaped laterals 0.5–1.5 cm long, arising from a basal stolon with rhizoids; on rock in medium to rough-water, from low-tide level to 36 m deep. Southern Australia generally (Victor Harbor to Marino). Approx. 10 other species.

11. *Codium muelleri,* CODIALES: Medium green, much-branched, 10–25 cm high, with branches 2.5–5.0 mm diam.; attached to rock by an expanded holdfast, 2–11 m deep under slight to moderate water movement. Southern Australia generally, common in Gulf St Vincent. Approx. 7 other species. See Fig. 3.

12. *Acetabularia (Polyphysa) peniculus,* DASYCLADALES: White (calcified) stems 2–6 cm high, bearing whorls of colourless hairs and an apical cluster of green club-shaped branches 2–5 mm long, containing minute spherical reproductive cysts, usually on old cockle shells just below low-tide. Southern Australia generally but confined to calm water in bays (Outer Harbor region). One other species.

1
2
3
4
5
6
7
8
9
10
11
12

Plate II

PHAEOPHYTA (brown algae)

1. *Sargassum lacerifolium,* FUCALES: Medium brown, with 3-sided stems, bearing leaves 4–10 cm long, undulate and often spiny, near the base, with fertile, much divided upper fronds (to 30 cm long) developing in spring and early summer, with spherical vesicles and 3-sided spinous fertile receptacles; 1–20 m deep on coasts of moderate wave action. South-eastern Australia (Victor Harbor to Marino). Approx. 12 other species.

2. *Cystophora botryocystis,* FUCALES: Medium to dark brown, with long flexuous branches (to 2 m long) bearing laterals with clusters of small vesicles below densely clustered terete receptacles 0.5–1.5 cm long; 2–20 m deep on coasts of slight to moderate water movement. Southern Australia (Victor Harbor to Outer Harbor). Approx. 10 other species. See Fig. 9.

3. *Scaberia agardhii,* FUCALES: Medium to dark brown, with long (to 2 m) branches from near the base, the upper branches 2–4 mm thick, with a wart like surface due to scales and vesicles; from low-tide level to 40 m deep under moderate to strong water movement. Southern Australia (Victor Harbor to Outer Harbor).

4. *Ecklonia radiata,* **kelp,** LAMINARIALES: Medium to dark brown, 0.5–2.0 m high, with a branched holdfast, cylindrical stem, and flat blade 2–8 cm broad bearing flat leafy laterals from its edge, sometimes spinous; on moderate to rough-water coasts, from low-tide level to 40 m deep. Southern Australia (Victor Harbor to Outer Harbor).

5. *Carpoglossum confluens,* FUCALES: Dark brown, branched in one plane, with branches basally constricted, flat, 5–15 mm broad and relatively thick, spotted with fertile chambers; 2–40 m deep on coasts of moderate to strong water movement. South-eastern Australia (Victor Harbor).

6. *Xiphophora chondrophylla,* FUCALES: Dark brown, robust, with forked branches, 6–20 cm high, branches 3–6 mm broad, branch ends spotted with fertile chambers, holdfast split; from low-tide level to 6 m deep. South-eastern Australia (Victor Harbor).

7. *Scytothalia dorycarpa,* FUCALES: Medium to dark brown, much branched (to 2 m long) in one plane (but often twisted), with branches 3–6 mm broad, bearing axillary fertile receptacles 10–15 mm long, broad below and tapering to a point; from low-tide level to 44 m deep on rough-water coasts. Southern Australia (Victor Harbor).

8. *Seirococcus axillaris,* FUCALES: Medium brown, much branched with long flat branches (to 2 m long) in one plane, 4–12 mm broad, bearing dense clusters of terete axillary receptacles 2–5 mm long and 0.5–1.0 mm diam.; from 3–40 m deep, on rough-water coasts. South-eastern Australia (Victor Harbor).

9. *Cystophora platylobium,* FUCALES: Medium brown, with long flat branches (to 2 m long) bearing from the edges flat pinnate laterals with occasional basal vesicles and flat elliptic-lanceolate receptacles 1–3 cm long; 1–48 m deep, on rough-water coasts. South-eastern Australia (Victor Harbor and Fleurieu Peninsula). Approx. 10 other species. See Fig. 2.

10. *Scytosiphon lomentaria,* SCYTOSIPHONALES: Medium to dark brown, unbranched, cylindrical or compressed, hollow, fronds 5–30 cm long; just above low-tide level. Cosmopolitan (Victor Harbor to Outer Harbor).

11. *Hormosira banksii,* **sea grapes,** FUCALES: Medium brown, with chains of spherical to elongate bladders (depending on degree of wave action) 1–3 cm in diam., spotted with fertile chambers containing the sex organs; in the lower third of the intertidal region. Southern Australia (Victor Harbor to St Kilda; the form depicted is from St Kilda).

12. *Acrocarpia paniculata,* FUCALES:Dark brown, very much irregularly branched, to 1.5 m long, with slender branches and terminal receptacles 1–4 mm long and 0.5–1.0 mm diam.; no vesicles; from low-tide level to 42 m deep. South-eastern Australia (Victor Harbor to Port Noarlunga). Fig. 12a shows detail of an upper branch.

1
2
3
4
5
6
7
8
9
10
11
12a
12

Plate III

RHODOPHYTA (red algae)

1. *Euptilota articulata,* CERAMIALES: Medium to dark red, 5–20 cm high, elegant, much-branched with prominent main axes bearing on 2 sides much-branched slender lateral branchlets; 2–20 m deep on rough-water coasts. Southern Australia (Victor Harbor).

2. *Areschougia congesta,* GIGARTINALES: Dark red to brown-red, 5–30 cm high, much-branched (to varying degrees) with cylindrical branches; 1–10 m deep on rough-water to moderate coasts. Southern Australia (Victor Harbor to Marino). Approx. 2 other species.

3. *Thamnoclonium dichotomum,* CRYPTONEMIALES: Dark brown-red, 5–15 cm high, with long slightly compressed branches covered with scales, divided near the base; 2–30 m deep, on rough-water to moderate coasts. South-eastern Australia from Ceduna east. (Victor Harbor to Marino).

4. *Botryocladia obovata,* RHODYMENIALES: Brown-red, 3–15 cm high, with branches covered with club-shaped vesicles (0.5–1.5 cm long) near their apices, divided below; on moderate to rough-water coasts, 1–10 m deep. Southern Australia (Victor Harbor to Outer Harbor).

5. *Callophycus laxus,* GIGARTINALES: Red-purple, 10–30 cm high, with laterals opposite, slightly compressed, 2–3 mm broad; 2–20 m deep on rough-water coasts. South-eastern Australia (Victor Harbor). 2 other species.

6. *Phacelocarpus labillardieri,* GIGARTINALES: Medium to dark red, 10–50 cm high, much branched with opposite feathery rows of short branchlets 2–4 mm long, divided near the base; 1–30 m deep on rough-water coasts. Southern Australia (Fleurieu Peninsula). Approx. 4 other species.

7. *Thuretia quercifolia,* CERAMIALES: Medium red, darker at the growing apices, 10–30 cm high, moderately branched and with a flat (1.0–1.5 cm broad) spongy network of slender filaments; 2–25 m deep on rough-water to moderate coasts. Southern Australia (Victor Harbor to Marino). 1 other species.

8. *Jania tenuissima,* CORALLINALES: Pale red, 1–5 cm high, epiphytic on the seagrass *Amphibolis,* with forked jointed calcified segments and reproductive conceptacles below branching; 1–5 m deep on coasts of moderate water movement. Southern Australia (Victor Harbor to Marino). Approx. 4 other species.

9. *Dasya extensa,* CERAMIALES: Medium to dark red, 10–50 cm high, much-branched, with all branches densely covered on all sides with slender coloured branched filaments; 2–25 m deep on rough-water to calm coasts. Southern Australia (Victor Harbor to Outer Harbor). Approx. 10 other species.

10. *Asparagopsis armata,* **Australian sea-rover,** BONNEMAISONIALES: Light red to grey-red, often bleached above, 10–20 cm high, much-branched with slender laterals, lower laterals with reflexed spines, upper with globular reproductive organs. The spinous laterals attach to other algae and this species has migrated to the Mediterranean and southern England, probably on ships' hulls; just below low-tide level on rough-water to moderate coasts. Southern Australia (Fleurieu Peninsula). 1 other species.

11. *Laurencia filiformis,* CERAMIALES: Medium to dark red, 5–20 cm high, much-branched, with irregular slender cylindrical branches which have slightly swollen apices; 1–20 m deep under moderate water movement. Southern Australia (Victor Harbor to Outer Harbor). Approx. 6 other species.

12. *Melanthalia obtusata,* GIGARTINALES: Dark red-purple, 10–40 cm high, branched largely in one plane with distant branches and rounded axils; branches compressed, 2–4 mm broad; on rough-water coasts, 1–15 m deep. South-eastern Australia (Victor Harbor). 1 other species.

1
2
3
4
5
6
7
8
9
10
11
12

Plate IV

FUNGI

1. **Lycoperdon depressum*, LYCOPERDACEAE: To 6 cm diam., spherical or pear-shaped, often tapering to a short stem, covered in small white pointed warts and granules, becoming yellowish to dirty brown when old. On lawns. All States except WA, NT and Qld; New Zealand; S. Africa and Europe.

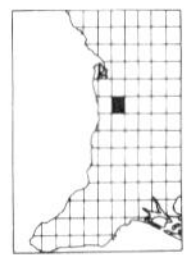

2. *Rhizopogon* sp., TREMELLACEAE: More or less spherical, to 5 cm diam., with white strands. In groups partially buried in soil and litter beneath pine trees. Probably introduced.

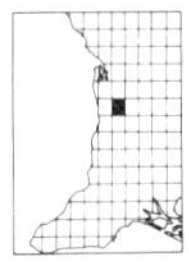

3. *Ramaria sinapicolor*, RAMARIACEAE: To 12 cm high, branched, with slender stems. In native forests on the ground. WA, NSW, Vic.

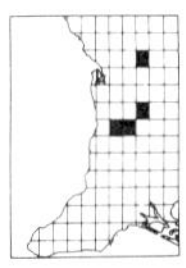

4. **Lycoperdon glabrescens*, LYCOPERDACEAE: To 5 cm diam., more or less spherical, tapering to a stem-like base, turning dark to purplish when old. On the ground in groups, usually in pastures. All States except WA, NT and Qld; ?Europe.

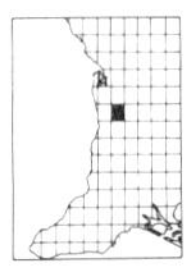

5. **Scleroderma flavidum*, SCLERODERMATACEAE: To 5 cm diam., spherical or pear-shaped, often lobed. On the ground in groups, usually under eucalypts. All mainland States except NT; ?Europe.

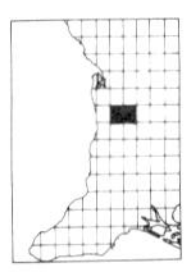

6. *Ramaria* sp., RAMARIACEAE: To 15 cm high and broad, resembling a cauliflower. On the ground in forests.

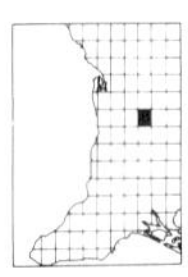

7. *Stemonitis* sp., STEMONITACEAE: To 1 cm high. Crowded, readily broken up by moving air. On decaying wood.

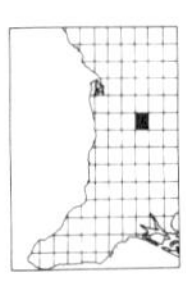

8. *Calocera sinensis*, DACRYMYCETACEAE: To 2 cm high, cylindric, sticky. Sometimes with several terminal branches, tips usually pointed and paler than the remainder of the fruiting body. Together in large numbers on rotting wood. All mainland States except NT.

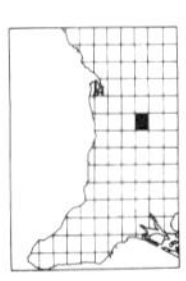

9. **Clavaria cristata*, CLAVARIACEAE: To 18 mm high, terminal branches usually pointed, short and forked. On rotting vegetation, associated with European trees. ?Europe.

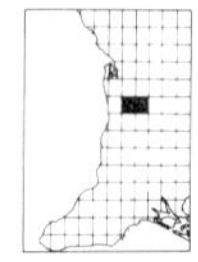

10. **Tremella fuciformis*, a **jelly fungus**, TREMELLACEAE: To 10 cm across, gelatinous. Often in groups on dead wood. ?Europe.

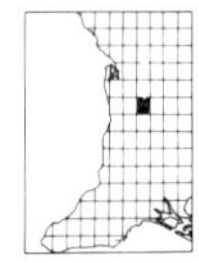

11. *Heterotextus flavus*, DACRYMYCETACEAE: 1–2 cm wide, cone-shaped, translucent, gelatinous with a convex outer surface when moist but flattening when dry. Often on wooden fences. Tas.

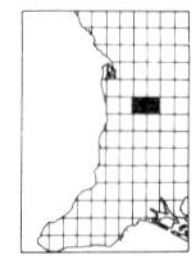

12. *Corticium* sp., CORTICIACEAE: 2–3 cm high, 5–20 mm wide. On dead wood.

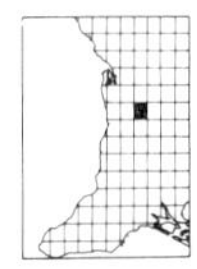

13. *Podoscypha petalodes*, PODOSCYPHACEAE: Stem to 9 cm high, 1–4 cm wide; tough, leathery. Grows at base of tree trunks and on buried roots. NSW, Vic.

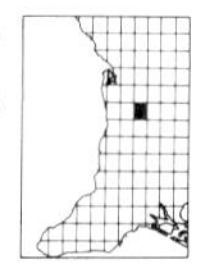

14. *Clavulinopsis amoena*, CLAVARIACEAE: To 4 cm high, segments to 7 mm thick, at first tapering at both ends, becoming branched, sometimes twisted. On the ground among leafy-liverworts and pines. NSW, Vic., Tas.

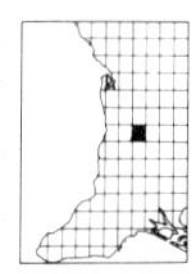

15. *Calostoma fuscum*, CALOSTOMATACEAE: To 3 cm diam., stem to 6 cm long; exposed cap has an opening with 4–7 raised scarlet rays. On clay soil. All States except NT.

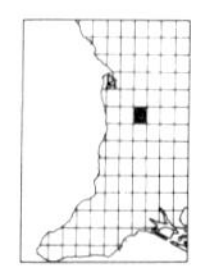

16. *Aseroe rubra*, CLATHRACEAE: 4–8 pairs of arms to 8 cm long, emerging from a purple 3 cm diam. egg-like stage. Often in groups, growing in native forest, woody or almond chips. Qld, NSW, Vic.

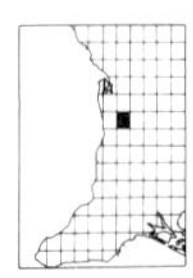

(*Continued*)

1
2
3
4
5
6
7
8
9
10
11
12
13
14
15
16

Plate V

FUNGI

(Continued)

1. **Trametes cinnabarina,* CORIOLACEAE: Fan-shaped leathery bracket attached by a broad base, to 10 cm long, to 6 cm wide, 1–2 cm thick; under surface convex, rough and hairy at first, becoming smooth later; under surface covered with small circular pores. Singly or in groups on dead wood, especially pines. All States except NT; ?Europe.

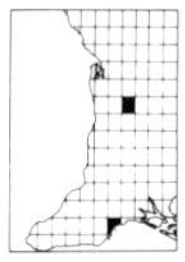

2. **Trametes versicolor,* CORIOLACEAE: Cap thin, overlapping fan-shaped leathery bracket, to 8 cm or more wide, to 5 mm thick and leathery; under surface smooth to velvety, with bands of various colours; edge of the brackets lobed and pale; under surface white to pale grey with small round pores. Scattered close together on fallen timber or on stumps. All States; New Zealand, Europe, Central America.

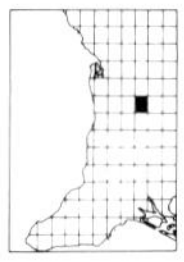

3. *Stereum illudens,* **Bracket fungus,** STEREACEAE: with edges upturned, to 10 cm wide; upper surface hairy towards the edges which are usually irregularly lobed; under surface smooth. Occurs as single or numerous brackets on dead wood. All States except NT, Qld and Tas.

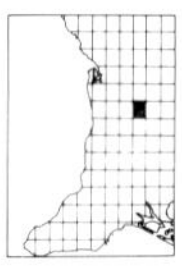

4. *Dictyopanus pusillum,* TRICHOLOMATACEAE: Cap 3–5 mm diam., edge incurving with pores below; stem lateral, 1–2 mm long. In large groups often on dead wood or trunks. All States except WA, NT and Tas.

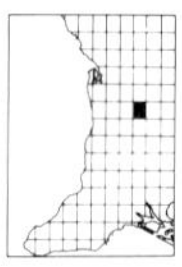

5. **Suillus luteus,* BOLETACEAE: Cap to 10 cm diam.; the skin slimy, readily peeling; flesh thick, soft, white to pale-yellow, spongy below; stem occasionally to 10 cm long, stout, solid, with a sticky ring. In groups on the ground under introduced pine trees. All States except NT, Qld and Tas; almost cosmopolitan.

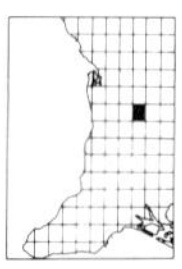

6. **Daedalia* sp., POLYPORACEAE: Irregular margin, no stem, with tube-like structures below. On buried stumps or rotting wood. Probably introduced.

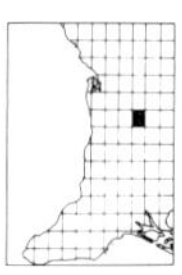

7. *Boletellus obscurecoccineus,* BOLETACEAE: Cap to 6 cm diam., spongy below; stem to 10 cm long. On the ground in native forests.

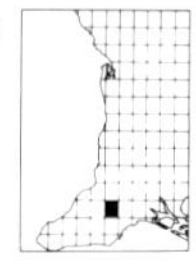

8. **Suillus granulatus,* BOLETACEAE: Cap to 10 cm diam., when young covered with a brownish slime which disappears when older leaving a dry skin that easily peels; flesh yellowish, soft and watery, spongy below; stem to 6 cm long and 1.5 cm thick, solid, encrusted with brown to black granules. In groups on the ground under introduced pines. All mainland States except NT; almost cosmopolitan.

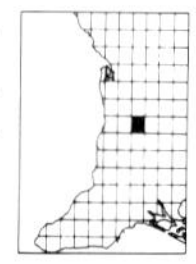

9. *Phellodon niger,* BANKERACEAE: Cap to 3 cm diam., tough, spongy below; stem c. 2.5 cm long, to 3 mm diam., smooth. Firmly attached to dead wood. NSW.

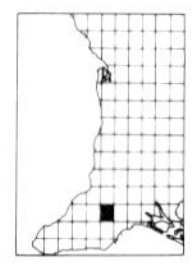

10. **Nidula emodensis,* a **bird nest fungus,** NIDULARIACEAE: Cup-like structure to 5 mm diam., with a flared margin, to 7 mm deep, interior of cup smooth and brown with smooth lentil-shaped bodies containing spores. In groups on dead wood. NSW, Vic.; ?Europe.

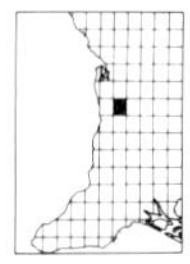

11. *Xerocomus fuscescens,* BOLETACEAE: Cap to 15 cm diam., convex to flat and irregular, smooth to velvety, slightly sticky when wet; flesh whitish becoming dark brown when cut, spongy below and turning dark brown when bruised; stem to 6 cm long and 4 cm thick, stout. On the ground under candle pines. NSW, Vic.

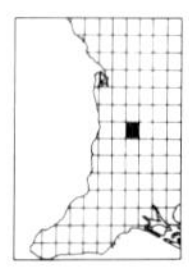

12. **Cyathus olla,* a **bird nest fungus,** NIDULARIACEAE: Cup-like structure to 12 mm diam., with flared margins, to 15 mm deep, covered in fine hairs, interior of cup smooth, shining, lead-coloured, with smooth lentil-shaped bodies containing spores. Growing in groups on rotting twigs and stems on the ground. NSW, Vic.; worldwide.

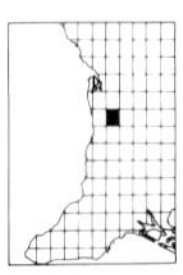

(Continued)

1
2
3
4
5
6
7
8
9
10
11
12

PLATE VI

FUNGI
(Continued)

1. **Crepidotus applanatus,* CREPIDOTACEAE: Cap to 5 cm diam., oval or kidney-shaped, soft and limp; the surface woolly at the point of attachment; gills crowded, white becoming cinnamon. On decaying wood in eucalypt and pine forests. NSW, Vic.; ?Europe.

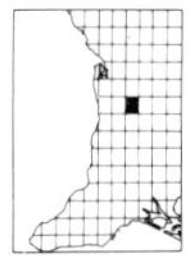

2. **Schizophyllum commune,* SCHIZOPHYLLACEAE: Cap to 5 cm across, shell-shaped or fan-shaped, tough and reviving after drying, the flesh brown when moist, attached by a narrow stem-like base. In groups on dead wood. All States; worldwide.

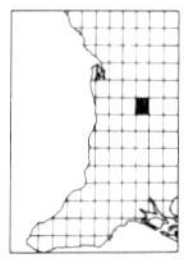

3. **Crepidotus variabilis,* CREPIDOTACEAE: Cap to 3 cm across, semi-circular to kidney-shaped, margin incurved; gills well-spaced, white becoming pinkish then pale cinnamon with age. On fallen branches and wood, single or in groups. NSW, Vic.; Europe.

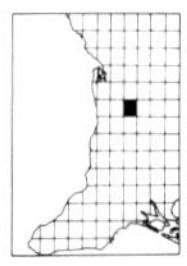

4. **Crepidotus* sp., CREPIDOTACEAE: Cap fan-shaped, to 5 cm diam., attached to one side with no definite stem; surface dry and woolly with an inturned margin; gills crowded. On trunks and fallen wood. ?Europe.

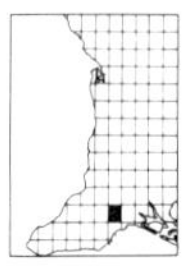

5. **Lentinellus* sp., LENTINELLACEAE: Cap to 8 cm diam.; stem lateral, to 5 cm long; gills crowded, pale brown becoming pinkish with age. Growing in compact clusters on wood. ?Europe.

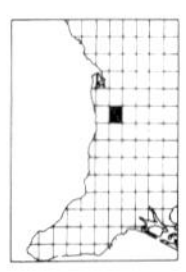

6. *Melanotus hepatochrous,* STROPHARIACEAE: Cap to 2.5 cm diam.; stem short, curved, to 3 mm long, with a fluffy base; gills crowded, many short. On trunks of eucalypts.

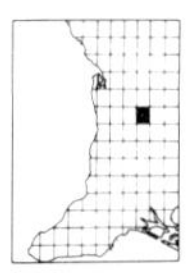

7. **Panaeolus* sp., COPRINACEAE: Cap to 4 cm diam.; stem to 6 cm long, slender, base slightly thickened, slightly hollow; gills crowded, many short, narrow, blackish with white margins. In grass. ?Europe.

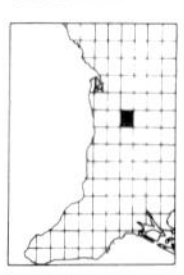

8. **Coprinus comatus,* **inky cap,** COPRINACEAE: Cap to 14 cm high and 6 cm diam.; stem to 15 cm long and 1.5 cm thick, hollow; gills free, broad, crowded, at first white then pink and finally black. On the ground among grass, along roadsides. All States except WA, NT and Tas; New Zealand, Europe, South Africa, the Americas.

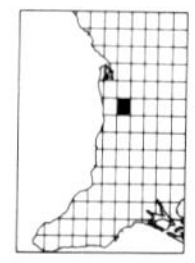

9. **Mycena subgalericulata,* TRICHOLOMATACEAE: Cap to 3 cm diam.; stem to 5 cm long, cylindric, smooth and shiny, with hairy rooting base; gills well spaced, white at first, later greyish. In dense clusters on old wood and on trunks of living trees. NSW, Vic.; ?Europe.

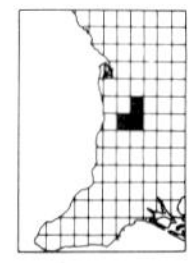

10. **Agrocybe* sp., BOLBITIACEAE: Cap to 5 cm diam.; stem to 6 cm long, slender, smooth, hollow; gills with short and longer gills mixed, pallid to crimson-brown. In gardens. ?Europe.

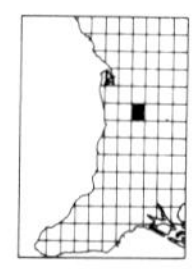

11. *Mycena subcapillaris,* TRICHOLOMATACEAE: Cap very small, to 5 mm diam., grooved; stem to 5 cm long, very slender, usually with a few scattered hairs; gills well-spaced, few. In groups on dead leaves and twigs and among moss. NSW, Vic.

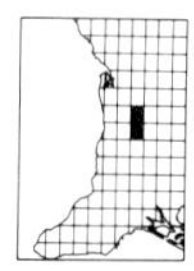

12. **Mycena* sp., TRICHOLOMATACEAE: Cap to 2 cm diam., lined; stem to 4 cm long, slender with a hairy rooting base; gills moderately crowded, white to brownish. In groups on rotting wood. ?Europe.

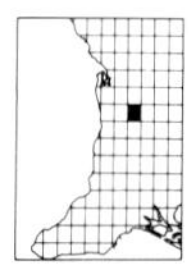

13. *Mycena subnigra,* TRICHOLOMATACEAE: Cap to 2 cm diam., faintly lined; stem to 5 cm long, slender, with a hairy rooting base; gills moderately crowded, swelling in the middle, greyish. In dense clumps on stumps.

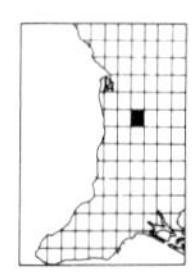

14. **Mycena* sp., TRICHOLOMATACEAE: Cap to 1 cm diam., lined; stem to 5 cm long, slender; gills well-spaced, white to greyish. On the ground. ?Europe.

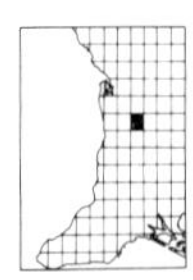

(Continued)

1
2
3
4
5
6
7
8
9
10
11
12
13
14

Plate VII

FUNGI
(Continued)

1. *Omphalina ericetorum,* TRICHOLOMATACEAE: Cap to 2 cm diam., margin finely grooved; stem to 2 cm long, slender, smooth; gills well-spaced, joined by veins, pale brown. In groups on the ground. Vic.

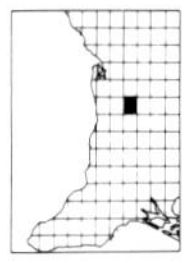

2. **Hypholoma fasciculare,* STROPHARIACEAE: Cap to 4 cm diam., with whitish veil fragments; stem to 5 cm long, to 1 cm thick, scaly; gills crowded, at first yellowish, then brownish-purple. In clumps, on old stumps and on buried wood. NSW, Vic.; Europe, N. America, S. Africa.

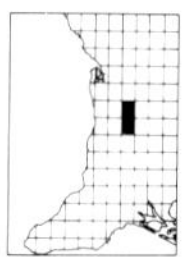

3. *Camarophyllus niveus,* HYGROPHORACEAE: Cap to 2 cm diam.; stem to 4 cm long, to 0.5 cm thick, slender, slightly hollow, silky; gills well-spaced, edges a little thick. On the ground under conifers. NSW, Vic; Europe.

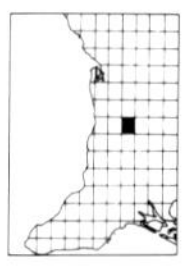

4. *Cortinarius sublargus,* CORTINARIACEAE: Cap to 9 cm diam., covered with small fibres and furrows; stem to 9 cm long, with a swollen base and conical root, to 2.3 cm thick, solid. In sandy soil often in recently burnt country.

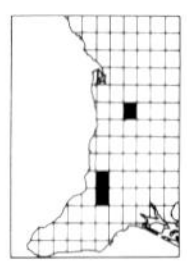

5. *Omphalina chromacea,* TRICHOLOMATACEAE: Cap to 2 cm diam.; stem to 2 cm long, smooth, solid; gills well-spaced, thick, yellow. In groups on bare ground or in low moss. WA, NSW, Vic.

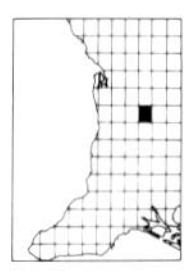

6. *Cortinarius subarcheri,* CORTINARIACEAE: Cap to 8.7 cm diam., sometimes irregular; stem to 3.7 cm long, at first thick, becoming more elongate and oval-shaped; gills slightly toothed, moderately crowded, pinkish-cinnamon tinged with violet. On leaf litter under eucalypt forests. All mainland States except NT.

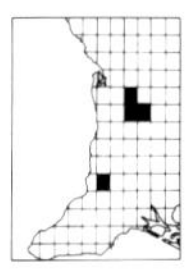

7. *Collybia elegans,* TRICHOLOMATACEAE: Cap to 5 cm diam., slightly velvety; stem to 6 cm long, slender and hollow; gills crowded, cream-coloured. In groups among leaf litter or on rotten wood. NSW, Vic.

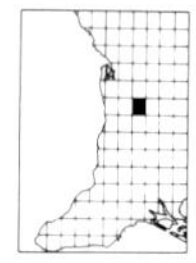

8. *Psilocybe subaeruginosa,* STROPHARIACEAE: Cap to 5 cm diam.; stem to 10 cm long, slender, slightly swollen and woolly at the base, a cobweb-like veil connecting it to the cap when young, often blotched and greenish-blue; gills crowded, smoky-brown becoming blackish. Solitary or in groups on soil and often on leaf litter under eucalypt forests. All mainland States except NT.

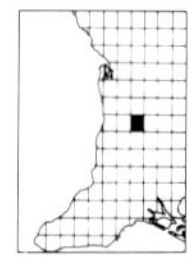

9. *Xerula radicata,* XERULACEAE: Cap to 7.5 cm diam.; stem to 20 cm long, tapering towards the top and with a long taproot; gills broad and rather thick, well-spaced, white. Solitary or in small groups on the ground at the base of old stumps and over buried wood. All mainland States except NT.

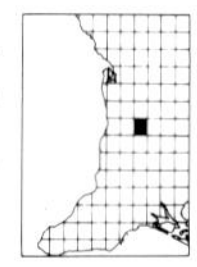

10. **Stropharia* sp., STROPHARIACEAE: Cap to 4 cm diam.; stem 8–9 cm long. On animal dung. All mainland States except NT; ?Europe.

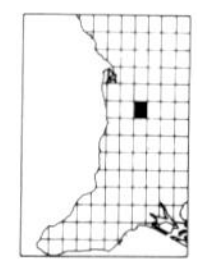

11. *Mycena vinacea,* TRICHOLOMATACEAE: Cap to 3 cm diam., bell-shaped; stem to 7 cm long, slender, smooth, hollow, woolly at the base; gills crowded, the same colour as the cap. In clusters on fallen wood, stumps and among leaf litter in forests. NSW, Vic.

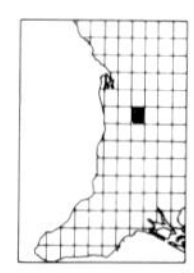

12. *Marasmius crinis-equi,* TRICHOLOMATACEAE: Cap to 4 mm diam., flesh very thin; stem 2.5–14.0 cm long, very slender, hair-like, smooth, tough; gills well-spaced, 8–11, whitish. In groups on plant debris in forests. Qld, NSW, Vic.

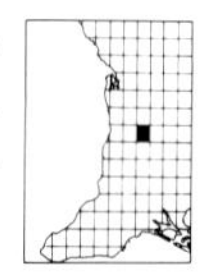

(Continued)

1
2
3
4
5
6
7
8
9
10
11
12

PLATE VIII

FUNGI
(Continued)

1. **Volvariella speciosa,* PLUTEACEAE: Cap to 12 cm diam., flesh soft, white and watery; stem to 20 cm long, to 2 cm thick, solid, easily detached from cap; gills white to pinky-brown, broad, crowded. Solitary often in manured soil, rotting grass. WA, NSW, Vic.; Europe, N. America, S. Africa.

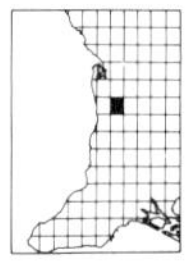

2. *Agaricus campestris,* **fold mushroom,** AGARICACEAE: Cap to 12 cm diam., silky, occasionally with a few small scales; stem to 8 cm long, to 4 cm thick, solid or slightly hollow; easily detachable from the cap, spreading ring above the middle of the stem; gills not attached to stem, whitish-pink to pink then purple-brown, crowded. On the ground in fields and along roadsides. All States; Europe, Asia, N. America, S. Africa.

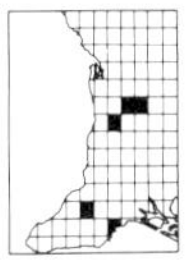

3. **Armillaria luteobubalina,* TRICHOLOMATACEAE: Cap to more than 10 cm diam.; stem to 20 cm long, usually darker at the base, often with fibres; gills pale, creamy-pink to cinnamon, crowded. In dense clumps at the base of trees and shrubs. All States except NT, Qld and Tas; ?Europe.

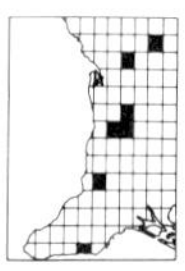

4. **Gymnopilus pampeanus,* CORTINARIACEAE: Cap often more than 20 cm diam., with small fibres and scales; stem c. 9 cm long, thick, and often swollen and curved at the base, same colour as cap; high on the stem a thin rusty-brown ring covered with spores; gills golden to rusty-brown, crowded, thin. In dense clumps at the base of trees and stumps. NSW, Vic.; ?Europe, ?N. America.

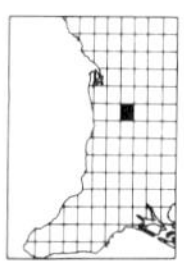

5. *Amanita xanthocephala,* AMANITACEAE: Cap to 3 cm diam., striate margin with yellow-whitish warts; stem to 4 cm long, to 1 cm thick, swollen at the base; gills white to lemon. Not attached to stem. Often solitary, on the ground in native forests. All States except NT, Qld and Tas.

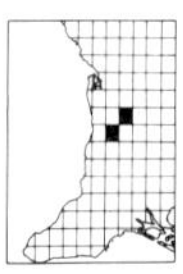

6. *Cortinarius archeri,* CORTINARIACEAE: Cap to 9 cm diam., very sticky and slimy when moist, violet at first, later becoming dry and tan; stem to 9 cm long, to 2.5 cm wide swollen at the base, very sticky, hollow in the lower portion; gills a violet tinge, becoming cinnamon-brown later, moderately crowded. On the ground in native forests. All mainland States except NT.

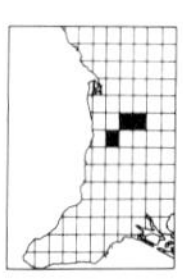

7. *Amanita grisea,* AMANITACEAE: Cap to 10 cm diam., slightly sticky with whitish fragments; stem to 13 cm long, 1–2 cm thick; gills cream-coloured, moderately crowded, deep. Not attached to stem. On the ground near *Eucalyptus rubida.* Tas.

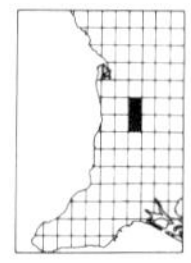

8. **Amanita muscaria,* fly agaric, AMANITACEAE: Cap often more than 15 cm diam., with white warts more concentrated in the centre; stem to 20 cm long, with a pendulous ring near the top and a swollen base with zones of white scales; gills white, not attached to stem. Under birch, pine and oak trees. NSW, Vic., Tas.; Europe.

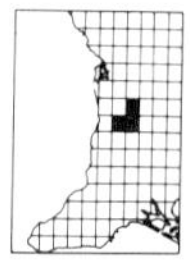

9. *Amanita* sp., AMANITACEAE: Cap to 8 cm diam., convex to flattened, stiff and brittle; stem to 8 cm long, brittle; gills crowded, white. On the ground near acacias.

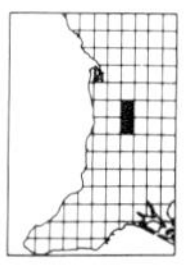

10. *Gymnopilus* sp., CORTINARIACEAE: Cap to 12 cm diam., with small fibres; stem tall often swollen and curved at the base, with a thin rusty-brown ring; gills golden to rusty-brown, crowded, thin. On stumps and dead wood. Probably introduced.

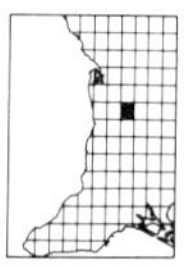

11. **Paxillus involutus,* PAXILLACEAE: Cap to 12 cm diam.; staining dark brown when damaged; stem to 6 cm long, to 1 cm thick, thicker at the top, smooth, darkening when handled; gills narrow, crowded, yellowish becoming olive-brown, and becoming dark-brown when bruised. Grows singly or in groups, sometimes forming 'fairy rings' on the ground among grass and in forest, also under oaks. NSW, Vic.; Europe, N. America, S. Africa.

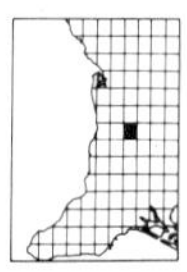

12. *Amanita grossa,* AMANITACEAE: Cap to 18 cm diam., covered with scattered warts which are often pointed; stem to 15 cm long, to 4.4 cm thick at base to 2.5 cm in the upper part; gills just reaching the stem, moderately crowded, creamy-white colour. On the ground in groups in native forest, has strong rancid or sour smell. NSW.

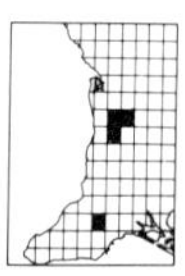

(Continued)

1
2
3
4
5
6
7
8
9
10
11
12

Plate IX

FUNGI

(*Continued*)

1. *Cortinarius australiensis,* CORTINARIACEAE: Cap very large, to 18 cm or more diam.; stem to 15 cm long, to 7 cm thick when mature, swollen below with a root-like base; gills attached to stem, crowded, narrow, pale-brown at first then rusty-brown. On the ground in eucalypt forests; mycorrhizal. NSW, Vic.

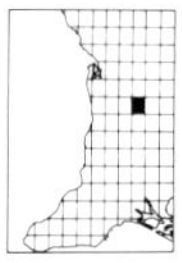

2. *Agaricus vinacea,* AGARICACEAE: Cap to 10 cm diam., covered with fine fibres; stem to 9 cm long, to 1.2 cm thick, small cottony scales below the ring, swollen at the base, solid to hollow; gills not attached to stem, crowded, pinkish to brown. Among debris under eucalypt forest.

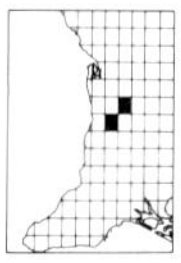

3. **Amanita rubescens,* AMANITACEAE: Cap to 10 cm diam., splitting, with greyish membranous warts; stem to 10 cm long, thick in the middle, to 2.5 cm diam., swollen below, staining red, ring soft and membranous; gills crowded, soft; white, spotted reddish-brown, staining red. In lawns under European trees. Europe, N. America, S. Africa.

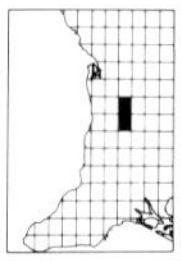

4. *Tricholoma coarctatum,* TRICHOLOMATACEAE: Cap to 6 cm diam., very irregular; stem c. 4 cm long, to 3 cm diam., stout, swollen in the middle, sometimes with a slight ring just below the gills; gills deep, crowded, cream, staining reddish-brown, with spots. In groups under eucalypts, partially buried in soil. Qld, NSW

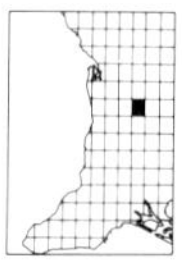

5. *Russula* sp., RUSSULACEAE: Cap to 9 cm diam., flesh solid, white; stem to 7 cm long, 3.7 cm thick, stout; gills moderately crowded, sometimes forking, pale buffy-white. In eucalypt forests.

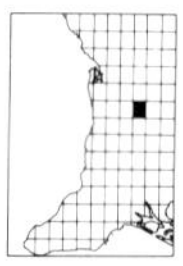

6. *Amanita* sp., AMANITACEAE: Cap to 10 cm diam.; stem to 15 cm long, 3-4 cm diam., stout, with a swollen base, solid ring very prominent; sometimes collapsing down along the stem; gills crowded, white to cream. Among grass and bracken, in native forests.

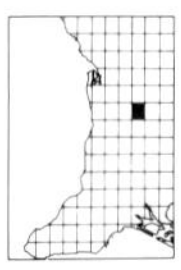

7. **Agaricus xanthodermus,* AGARICACEAE: Cap to 12 cm diam., staining yellow when bruised; stem to 12 cm long, slender, to 3 cm thick, with a swollen base, hollow, stains yellow when bruised, ring often stained yellow at the margin; gills whitish to greyish-pink to brownish-black. On the ground, on lawns, in woods and open country, under almond trees. All States except WA, NT and Tas; Europe, S. Africa.

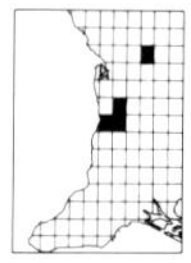

8. *Russula purpureoflava,* RUSSULACEAE: Cap to 6 cm diam., slightly sticky; stem to 4 cm long, to 1 cm diam., thick above, tapering downwards, occasionally with rosy tints; gills moderately crowded, rather narrow, tapering both ways. In native forests on the ground. NSW, Vic.

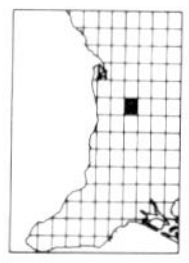

9. *Tricholoma sublilacinum,* TRICHOLOMATACEAE: Cap to 8 cm diam., smooth, lilac tinged when wet, whitish when dry; stem to 6 cm long, to 1 cm thick, solid, smooth; gills crowded, grey with a lilac tinge. Among grass. NSW, Vic.

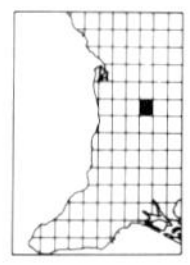

10. **Pluteus cervinus,* PLUTEACEAE: Cap to 11 cm diam., with a central knob, radially filamentous and streaked, sticky when wet; stem to 10 cm long, to 1.5 cm thick, slightly swollen at the base, with dark lines; gills not attached to stem, rounded next to the stem, crowded, white to flesh-pink to salmon-pink. Occurs singly or in groups on decayed wood on humus. All States except NT, Qld and Tas; Europe, N. America.

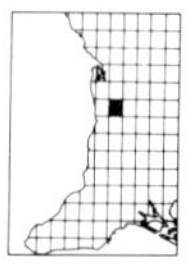

11. *Clitocybe eucalyptorum,* TRICHOLOMATACEAE: Cap to 15 cm diam., centre depressed when mature; edge wavy; stem to 10 cm long, to 2 cm thick, hollow; gills crowded, greyish to cream. On the ground in native forests. NSW, Vic.

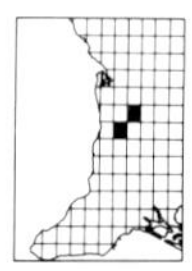

12. *Hygrophorus* sp., HYGROPHORACEAE: Cap to 5 cm diam., fragile, with a wavy edge, often folded; stem to 6 cm long, straight or furrowed; gills bright-yellow. Growing singly in leaf litter under pines. Probably introduced.

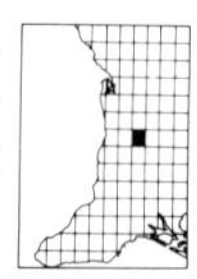

(*Continued*)

1
2
3
4
5
6
7
8
9
10
11
12

Plate X

FUNGI
(Continued)

1. *Psilocybe* sp., STROPHARIACEAE: Cap to 3 cm diam.; stem to 4 cm long, slender, slightly swollen at the base, to 3 mm thick; gills crowded, yellow-brown. On the ground. Probably introduced.

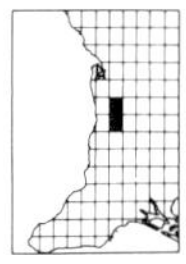

2. *Cortinarius* sp., CORTINARIACEAE: Cap to 7 cm diam.; stem to 7 cm long, hollow, dry with remains of cobweb-like veil attached, to 2 cm thick; gills crowded, cinnamon-brown to tawny. On the ground in native forest.

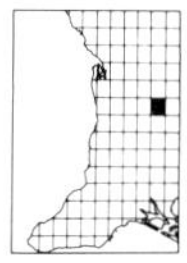

3. *Bolbitius vitellinus,* BOLBITIACEAE: Cap to 5 cm diam., slimy, grooved and often split, becoming paler with age; stem to 10 cm long, to 5 mm thick, tapering upward; rather fragile, hollow; gills not attached to stem, pale ochre to rusty-cinnamon with age, crowded. Among grass. NSW, Vic.

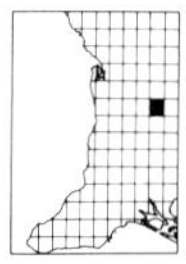

4. *Cortinarius sanguineus,* CORTINARIACEAE: Cap to 4 cm diam., smooth with edges sometimes slightly grooved; stem to 9 cm long, moderately stout, with fibres; gills crowded, dark-red later becoming rusty from the spores. On the ground in native forests. NSW, Vic.

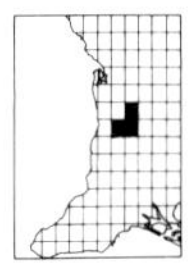

5. **Stropharia ferri,* STROPHARIACEAE: Cap to 10 cm diam.; stem to 10 cm high, to 3 cm thick, smooth, dry, hollow; gills crowded, smoky-grey at first, finally black, with white edges. On wood chips, under oaks. NSW; N. America and ?Europe.

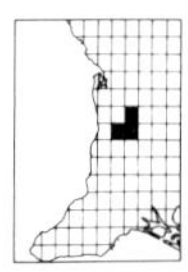

6. *Russula mariae,* RUSSULACEAE: Cap to 8 cm diam., flat, variable in colour; stem to 6 cm long, to 2 cm wide, stout, cylindrical, brittle, irregularly hollow; gills crowded, white to pale cream, occasionally forked. On ground in native forests. All mainland States except NT.

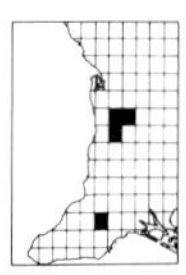

7. *Cortinarius lavendulensis,* CORTINARIACEAE: Cap to 9 cm diam.; stem to 8 cm long, to 2 cm thick, base usually swollen, with fibres, solid, or occasionally hollow; gills moderately crowded, pale buff to pale brown with a slight violet tint. Under bushes.

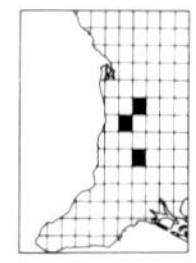

8. **Pholiota* sp., STROPHARIACEAE: Cap to 7 cm diam., with fine fibres when dry; stem to 9 cm long, and 1 cm thick, fine scales along the length, darker at the base; gills crowded, finely toothed, pallid-brown to olive-brown. On the ground in native forests. Probably introduced.

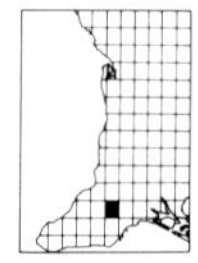

9. *Russula persanguinea,* RUSSULACEAE: Cap to 10 cm diam.; stem to 5 cm long, stout, smooth to slightly flattened, solid or sometimes hollow; gills equal and uniform in shape, rarely forked, crowded, pure white. On the ground.

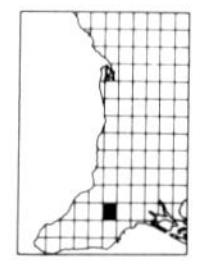

10. *Cortinarius cinnamoneus,* CORTINARIACEAE: Cap to 10 cm diam., silky or with small scales; stem to 9 cm long, to 1 cm thick, hollow, with fibres, slightly swollen; gills narrow, crowded, yellow at first then cinnamon. On the ground in native forests. NSW, Vic.

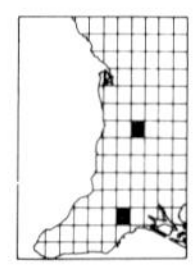

11. **Laccaria laccata,* TRICHOLOMATIACEAE: Cap to 6 cm diam.; stem to 8 cm long, to 1 cm thick, often curved or twisted, fibrous and tough, with small fibres and scales; gills thick and well-spaced, with intermediate short gills; flesh-pink to reddish-brown, with a powdery bloom when older. Common, usually on sandy loam. All States except WA and NT; Europe, N. America, S. Africa.

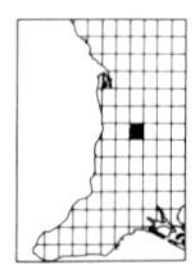

12. *Dermocybe austrovenetus,* CORTINARIACEAE: Cap to 8 cm diam.; stem to 8 cm long, rather stout, hollow, with fibres; gills crowded, ochre-coloured to rusty-brown. In groups on the ground in native forests. NSW, Vic.

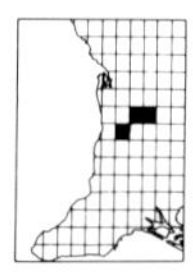

1
2
3
4
5
6
7
8
9
10
11
12

Plate XI

LICHENS

1. *Buellia subalbula,* PHYSCIACEAE: Forms a thin pale-grey layer mainly on limestone; fruiting bodies black, c. 1 mm across. NW, NU, EP, MU; WA, NSW, Vic., Tas.

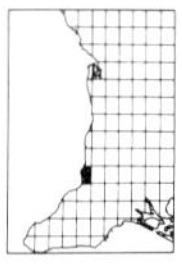

2. *Flavoparmelia jelinekii,* PARMELIACEAE: The leaf-like plant can be green or yellow-green and to 20 cm across; the disk-like fruiting bodies to 7 mm across; often grows on dead wood. NSW, Vic., Tas. *F. rutidota* has also been found.

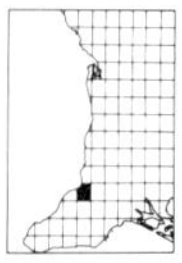

3. *Fulgensia subbracteata,* TELOSCHISTACEAE: The plant can be white to orangish and 1–2 cm diam.; the disk-like fruiting bodies reddish; grows on soil. NW, NU, GT, EA, EP, MU, YP, KI; WA, NSW, Vic., Tas.

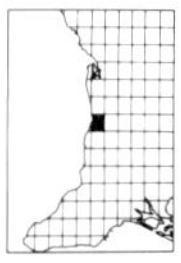

4. *Eremastrella crystallifera,* LECIDEACEAE: Forms a scaly grey to brown layer on soil; the disk-like fruiting bodies to c. 3 mm across. NU, GT, EA, EP, NL, MU, YP; WA, NSW, Vic., Tas.; N. America, Europe, Asia and Africa.

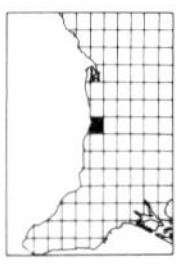

5. *Lichina confinis,* LICHINACEAE: Plant olive-green to black, to 1 cm tall; fruiting bodies 0.3–0.5 mm diam; grows on rocks near the sea. Vic., Tas.; N. America.

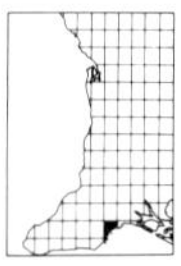

6. *Neofuscelia pulla,* PARMELIACEAE: The leaf-like plant olive-brown to yellowish-brown; the disk-like fruiting bodies to 8 mm across; grows on soil and rocks. NU, FR, EA, EP, NL, MU, KI; all States except NT.

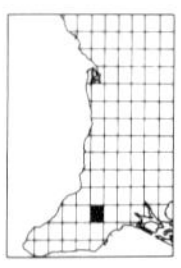

7. *Xanthoparmelia pseudohypoleia,* PARMELIACEAE: The leaf-like plant yellow-green, 5–8 cm across, often black-margined; the disk-like fruiting bodies 3–6 mm across; grows on rocks. Qld, NSW, Vic., Tas. 20–30 other species recorded. See Plate XII, Fig. 9.

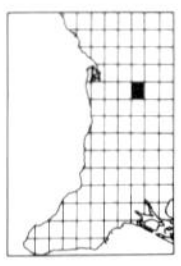

8. *Parmotrema chinense,* PARMELIACEAE: The leaf-like plant always blue-grey, to 15 cm across; the disk-like fruiting bodies very rare, to 7 mm across; note the fringing hairs; grows on trees. SE; WA, Qld, NSW, Vic.

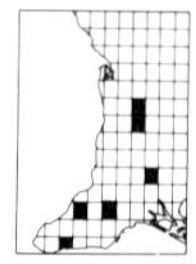

9. *Parmelina pruinata,* PARMELIACEAE: The leaf-like plant greyish, to at least 3 cm long; the disk-like fruiting bodies to 2.5 mm across; grows on trees and bushes. Distribution not available. WA, NSW, Vic., Tas.

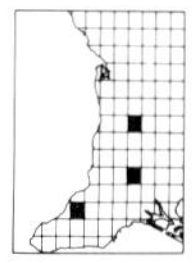

10. *Pseudocyphellaria neglecta,* LOBARIACEAE: The leaf-like plant always brown, to 12 cm across, yellow-pitted below; fruiting bodies to 2 mm across; grows in wet places on rocks. NSW, Vic., Tas.

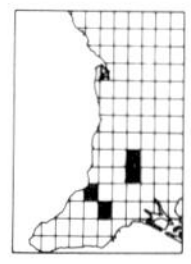

11. *Psora decipiens,* LECIDEACEAE: The scale-like plant pinkish, to 6 mm across; fruiting bodies (not shown) black, usually along the margins; grows on soil. NU, EA, EP, MU, YP; all States except Qld; N. America, Europe, Asia and Africa.

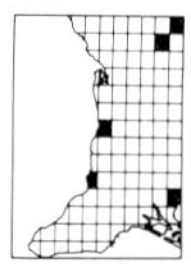

12. *Xanthoria ligulata,* TELOSCHISTACEAE: Plant to 8 cm across; fruiting bodies disk-like, to 2.5 mm across; grows on rocks and trees. NU, EP, KI; WA, NSW, Vic., Tas. *X. parietina,* which may be an introduction, has thinner wrinkled lobes.

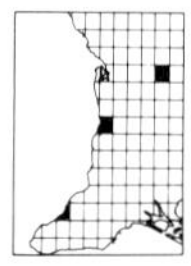

(Continued)

1
2
3
4
5
6
7
8
9
10
11
12

Plate XII

LICHENS
(Continued)

1. *Cladia ferdinandii,* CLADONIACEAE: Lace-like plant to 10 cm tall; fruiting bodies minute; grows on soil. EP, KI, SE; WA.

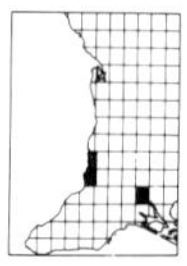

2. *Cladia aggregata,* CLADONIACEAE: Plant to 8 cm tall; fruiting bodies minute; grows on soil. FR, EP, MU, YP, KI, SE; all States except NT. *C. schizopora* is to only 3 cm tall.

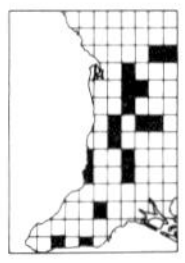

3. *Cladonia cervicornis,* CLADONIACEAE: Plant to 14 cm tall; fruiting bodies brown, to 3 mm across; usually grows on soil. FR, EP, NL, YP, KI, SE; WA, NSW, Vic., Tas. Several other species.

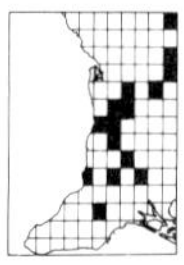

4. *Heterodea muelleri,* HETERODEACEAE: The leaf-like plant to 10 cm across; the marginal fruiting bodies to 1 mm across. NW, FR, EP, NL, MU, SE; all States.

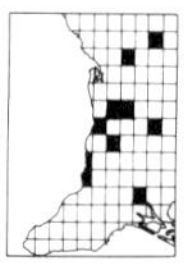

5. *Ramalina celastri,* RAMALINACEAE: Plants to 8 cm tall; the disk-like fruiting bodies 2–3 mm across; grows on trees and bushes. GT, EP, MU, YP, KI, SE; all States except NT. 3 other species.

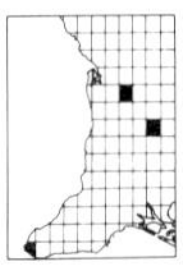

6. *Teloschistes chrysophthalmus,* TELOSCHISTACEAE: Plant orange to greyish-orange, to 2 cm across; the disk-like fruiting bodies to 6 mm across; grows on trees and bushes. EP, MU, YP, KI, SE; all States except NT. 3 other species.

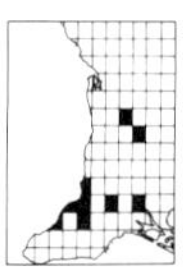

7. *Usnea inermis,* USNEACEAE: Plants 1–5 cm tall; disk-like fruiting bodies not developed; grows on trees and bushes. SE; all States except NT.

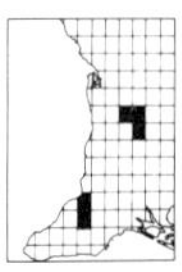

8. *Usnea scabrida,* USNEACEAE: Plant 1–5 cm tall; disk-like fruiting bodies to 1 cm across; grows on trees and bushes. EP, MU, YP; all States except NT. 4 other species.

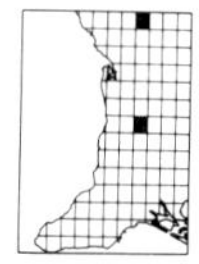

9. *Xanthoparmelia tasmanica,* PARMELIACEAE: The leaf-like plant yellow-green or blackening, to 15 cm across, brown-margined or black-margined; the disk-like fruiting bodies to 15 mm across; usually grows on rocks. NW, FR, EP, NL, MU, KI, SE; all States except NT; New Zealand, N. America and Africa. 20–30 other species. See Plate XI, Fig. 7.

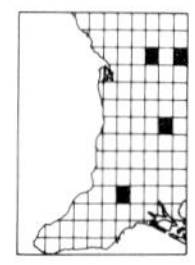

LIVERWORTS

10. *Cephaloziella exiliflora,* CEPHALOZIELLACEAE: Lvs c. 0.1 mm long, the upper whitish ones enlarged to c. 0.5 mm; often grows on burnt wood. FR, KI, SE; WA, Vic., Tas. Also *C. arctica*.

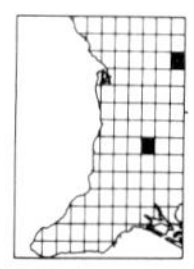

11. *Metzgeria decipiens,* METZGERIACEAE: Plant lobes c. 1 mm across; capsule less than 1 mm across; grows on tree trunks. NSW, Vic., Tas.; the Pacific.

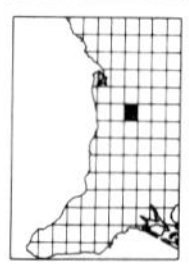

12. *Fossombronia* sp., CODONIACEAE: Several species, with lvs 0.5–5 mm broad; capsules usually c. 1 mm across; growing mainly on soils. FR, MU, KI, SE; all States except NT; worldwide.

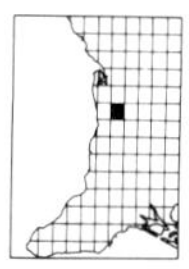

(Continued)

1
2
3
4
5
6
7
8
9
10
11
12

Plate XIII

LIVERWORTS
(*Continued*)

1. *Lophocolea semiteres,* GEOCALYCACEAE: Lvs c. 1–1.5 mm long; capsule c. 1 mm long. Most commonly grows on fallen logs. EP, KI, SE; all States except NT. 2 other species.

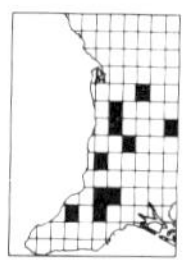

2. *Targionia lorbeeriana,* TARGIONIACEAE: Lobes of the plant 2–4 mm wide; capsules c. 2 mm long. Grows on soil. FR, EP, MU; WA, NSW, Vic., Tas.; New Zealand, Europe, Africa.

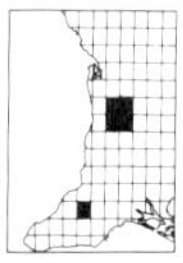

3. *Asterella drummondii,* AYTONIACEAE: Lobes of the plant 5–7 mm wide; capsules formed in heads 3–4 mm long. Grows on soil. FR, MU, KI, SE; WA, Qld, NSW, Vic.

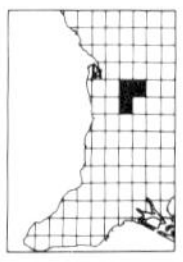

4. *Anthoceros laevis,* ANTHOCEROTACEAE: Lobes of the plant to 5 mm or more wide; note darkish colour; cylindrical fruiting structure to 2 cm long. Grows on soil. FR, EP, KI, SE; all States; cosmopolitan.

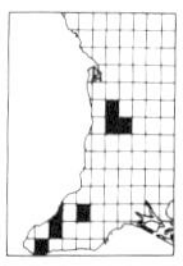

5. *Lunularia cruciata,* MARCHANTIACEAE: Lobes of the plant to c. 1 cm wide; note the crescent-shaped cups on the surface; capsules in heads to c. 4 mm across. Grows on soil and is common in gardens. MU, KI, SE; all States except NT; cosmopolitan.

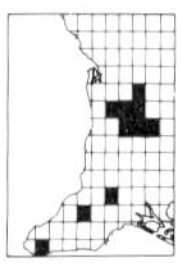

6. *Riccia* sp., RICCIACEAE: A large genus in which the lobes of the plant are c. 1–6 mm wide; the capsules are concealed inside the plant. Grows on soil. All regions except NU; all States; cosmopolitan.

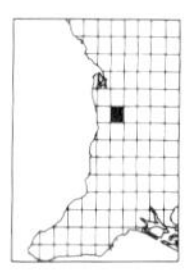

MOSSES

7. *Triquetrella papillata,* POTTIACEAE: Leafy stems 2–4 cm long; lvs in 3 rows, 1–2 mm long; capsules c. 1.5–2.0 mm long, on a somewhat wavy stem. GT, FR, EP, NL, MU, YP, KI, SE; all States except NT; New Zealand.

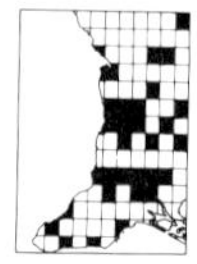

8. *Campylopus introflexus,* DICRANACEAE: Leafy stems 0.5–3 cm tall; lvs 4–6 mm long, usually ending in a bent bristle; capsules c. 2 mm long, on a wavy stem, furrowed when dry. FR, EP, NL, MU, KI, SE; all States except NT; New Zealand, Europe, the Pacific, the Americas. 2 other species.

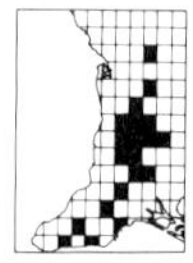

9. *Bartramia stricta,* BARTRAMIACEAE: Leafy stems 0.5–1.5 cm tall; lvs 1.5–2.5 mm long, narrowly pointed; capsules 1.0–1.5 mm long, on a straight stem. ?FR, ?MU; all States except NT; Europe, Asia, Africa, the Americas. 4 other species.

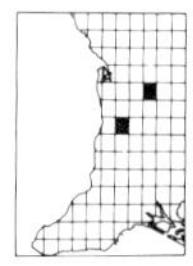

10. *Thuidium furfurosum,* THUIDIACEAE: Leafy stems creeping, to 10 cm long; lvs 1.5–2.0 mm long; capsules 2.0–2.5 mm long. NU, FR, EP, MU, YP, KI, SE; all States except NT; New Zealand, S. America.

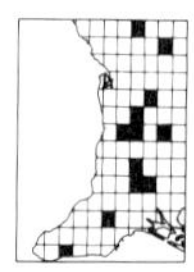

11. *Hypnum cupressiforme,* HYPNACEAE: Robust leafy stems 5–10 cm long, partly prostrate; lvs to 2.75 mm long; capsules 1–2 mm long. FR, KI, SE; all States except NT; cosmopolitan. Used in horticulture for its water-retaining properties.

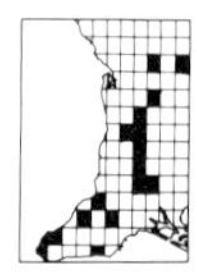

12. *Barbula calycina,* POTTIACEAE: Leafy stems 0.5–3.0 cm tall; lvs 2–3 mm long; capsules 1.25–2.0 mm long, on a long often wavy stem. NU, FR, EP, NL, MU, YP, KI, SE; all States except NT; New Zealand, Chile.

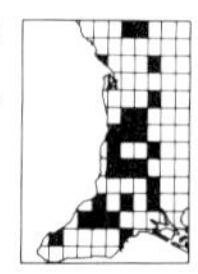

(*Continued*)

1
2
3
4
5
6
7
8
9
10
11
12

Plate XIV

MOSSES
(Continued)

1. *Polytrichum juniperinum,* POLYTRICHACEAE: Leafy stems 2–10 cm tall; lvs 5–8 mm long, with a terminal bristle; capsules 2.5–3.5 mm, with a hairy cap. MU, KI, SE; Qld, NSW, Vic., Tas.; almost cosmopolitan.

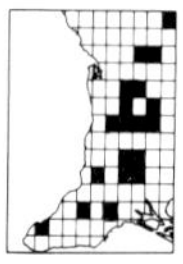

2. *Ceratodon purpureus,* DITRICHACEAE: Leafy stems 5–15 mm tall; lvs 1–2 mm long; capsules 2–3 mm long, squarish in section, with a minute swelling at the base, common after fires. NU, MU, KI, SE; all States except ?Qld; cosmopolitan.

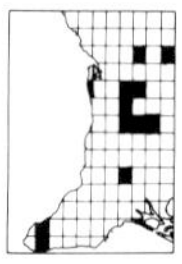

3. *Bryum torquescens,* BRYACEAE: Leafy stems 1–2.5 cm tall; lvs usually 1.5–4.0 mm long; capsules hanging down, usually 3–5 mm long. Vic.; cosmopolitan.

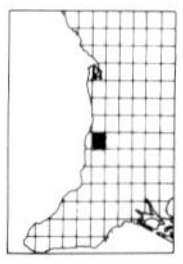

4. *Bryum argenteum,* BRYACEAE: Leafy stems usually less than 1.5 cm tall; lvs c. 1 mm long, silvery; capsule to 1.5 mm long. NW, FR, NL, MU, YP, SE; all States; cosmopolitan.

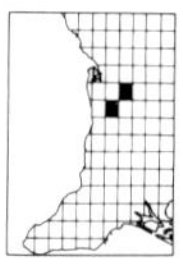

5. *Bryum* sp., BRYACEAE: Leafy stems usually 0.5–1 cm tall; lvs usually 0.6–1 mm long; capsules hanging down, usually c. 2 mm long. Close to *B. pachytheca.* Approx. 15 other species recorded.

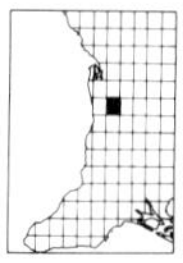

6. *Pottia davalliana,* POTTIACEAE: Leafy stems to 3 mm tall; lvs c. 1–2 mm long; capsule lacks teeth (remove cap), c. 1 mm long, erect. On disturbed soils. ?NSW, Vic.; Europe, N. Africa.

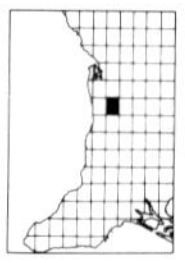

7. *Rhacopilum convolutaceum,* RHACOPILACEAE: Leafy stems creeping, to 1.5 cm long; lvs in 4 rows, the upper 2 rows smaller, 0.75–2.0 mm long, with an apical bristle; capsules 2.5–4.0 mm long. FR, EP, NL, YP, KI, SE; all States except NT; New Zealand.

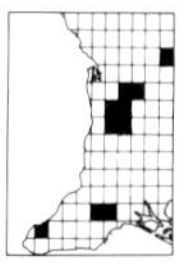

8. *Fissidens pungens,* FISSIDENTACEAE: Leafy stems weak, 2–4 mm long; lvs c. 1–2 mm long, in 1 plane; capsules to 1 mm long. NL, KI, SE; all States; New Zealand. Approx 9 other species recorded.

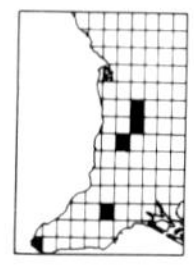

9. *Funaria hygrometrica,* FUNARIACEAE: Leafy stems 0.5–2.5 mm tall; lvs 2–4 mm long; capsules erect or hanging down, c. 2 mm long. All States; cosmopolitan. Approx. 6 other species recorded.

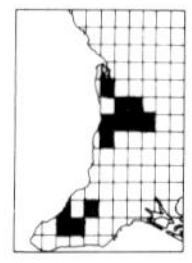

10. *Tortula papillosa,* POTTIACEAE: Leafy stems usually 7–8 mm tall; lvs with long apical bristles and small globular reproductive bodies on their surface, c. 3–4 mm long; capsules 2.0–2.5 mm long. FR, NL, MU, YP, KI, SE; WA, Qld, NSW, Vic., Tas.; almost cosmopolitan. 4 other species recorded.

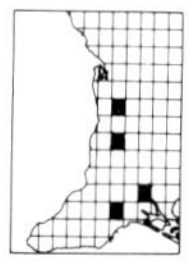

11. *Aloina ambigua,* POTTIACEAE: Leafy stems to 2 mm tall; lvs with broad infolded margins, 1–3 mm long; capsules 3.0–3.5 mm long. SE; Vic.; almost cosmopolitan. *A. sullivaniana* has long fine tips to the lvs.

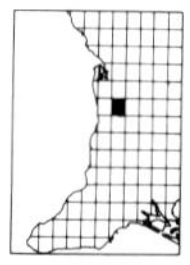

12. *Rhynchostegium tenuifolium,* BRACHYTHECIACEAE: Leafy stems prostrate, to 8 cm long; lvs 1.5–3.0 mm long; capsules c. 1 mm long. MU, KI, SE; all States except NT; New Zealand; S. America.

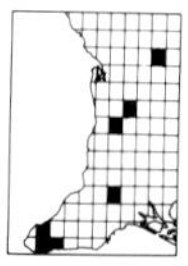

1
2
3
4
5
6
7
8
9
10
11
12

PLATE XV

OSMUNDACEAE

1. *Todea barbara*, **king fern**: Stem erect, to 30 cm tall; fronds to 2 m long; sporangia (1a) exposed on lower parts of fronds. Occurs in wet places. ?KI; Qld, NSW, Vic., Tas.; New Zealand, S. Africa.

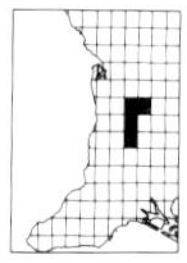

GLEICHENIACEAE

2. *Gleichenia microphylla*, coral fern: Fronds sprawling, repeatedly forked, 1–2 m long; sporangia (2a) exposed. Occurs in swamps or along streams. KI, SE; all States except NT; New Zealand, Malaya.

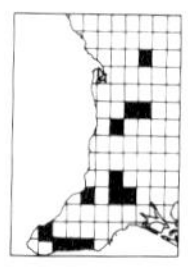

ADIANTACEAE

3. *Adiantum aethiopicum*, **common maiden-hair fern:** Fronds usually to 30 cm tall; sporangia (3a) in notches of the fronds, covered by a flap of tissue. Occurs on damp clay banks or among rocks. FR, EP, KI, SE; Qld, NSW, Vic., Tas.; New Zealand, Africa.

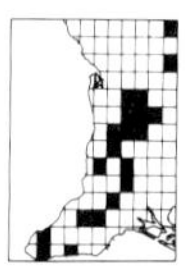

4. *Anogramma leptophylla*, **annual fern**: Fronds to 10 cm tall; sporangia (4a) along veins. Occurs on damp banks among grasses or rocky crevices. FR, EP, SE; WA, NSW, Vic., Tas.; temperate parts of the world.

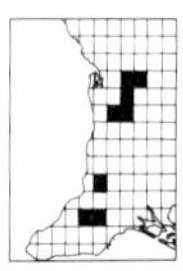

5. *Cheilanthes austrotenuifolia*, **rock fern**: Fronds to 45 cm tall, to 16 cm wide; sporangia (5a) protected by the reflexed frond margins. Occurs in moist regions, on banks and along streams. FR, EP, NL, MU, YP, KI, SE; all States. *C. distans*, **bristly cloak-fern**, fronds are densely hairy-scaly below when young.

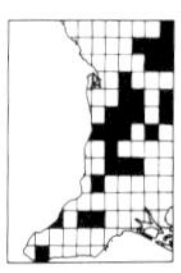

DENNSTAEDTIACEAE

6. *Hypolepis rugosula*, **ruddy ground-fern**: Fronds to 1 m tall; sporangia (6a) exposed or partially protected by a flap of tissue. KI, SE; WA, Qld, NSW, Vic., Tas. **H. dicksonioides*, **downy ground-fern**, has whitish-hairy (not pale brown) young fronds.

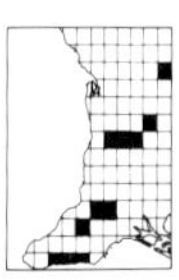

7. *Pteridium esculentum*, **bracken**: Fronds stiff, 0.3–2.0 m tall; sporangia (7a) along the margins, protected by 2 flaps. EP, KI, SE; Qld, NSW, Vic., Tas.; New Zealand, Pacific.

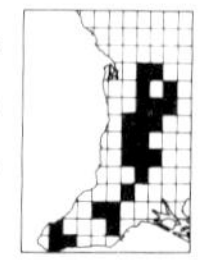

ASPLENIACEAE

8. *Asplenium flabellifolium*, **necklace fern**: Fronds to 25 cm long; sporangia (8a) along veins; often rooting at its tip. FR, NL, SE; WA, Qld, NSW, Vic., Tas.; New Zealand.

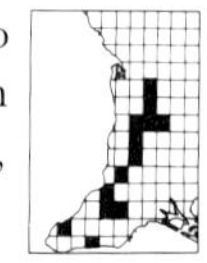

9. *Pleurosorus rutifolius*, **blanket fern**: Fronds rarely more than 12 cm tall, with dense whitish hairs; sporangia (9a) in elongate groups along veins. NW, LE, NU, GT, FR, EA, EP, NL, MU, KI, SE; all mainland States; New Zealand.

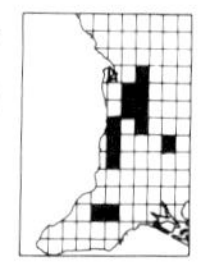

BLECHNACEAE

10. *Blechnum minus*, **soft water-fern**: Fronds to 1 m or more tall, sterile (10a) or fertile (10b); sporangia covered by turned-under frond margins. SE; Qld, NSW, Vic., Tas.; New Zealand. *B. wattsii*, **hard water-fern**, also has narrow-based pinnules but they do not get smaller towards the frond base.

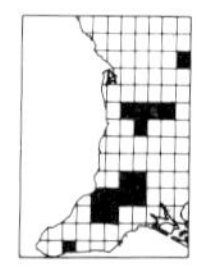

11. *Blechnum nudum*, **fishbone water-fern**: Short trunk sometimes present; fronds to 1 m tall; sporangia (11a) covered by turned-under frond margin. KI; Qld, NSW, Vic., Tas.

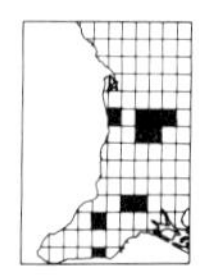

LINDSAEACEAE

12. *Lindsaea linearis*, **screw-fern**: Sterile fronds 6–10 cm tall; sori (12a) close to the margin, covered by a flap. KI, SE; all States; New Zealand, New Caledonia.

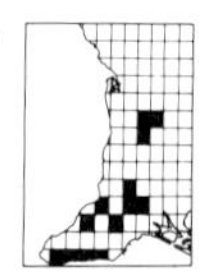

1
1a
2
2a
3
3a
4
4a
5
5a
6
6a
7
7a
8
8a
9
9a
10
10a
10b
11
11a
12
12a

Plate XVI

LYCOPODIACEAE

1. *Lycopodium laterale*, **slender clubmoss**: Perennial herb usually 10-45 cm tall; lvs 4-7 mm long; cones (1a) 8-10 mm long. KI; all States except NT; New Zealand, New Caledonia. 2 other species.

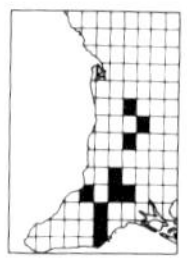

ISOETACEAE

2. *Isoetes drummondii*, **plain quillwort**: Perennial herb; rootstock 2-lobed or 3-lobed; lvs 5-22 cm long; sporangia in the base of the lvs large, dark brown. KI, SE; WA, Vic., Tas.

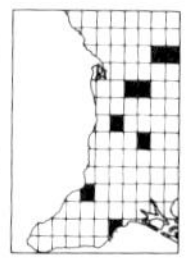

OPHIOGLOSSACEAE

3. *Ophioglossum lusitanicum*, **austral adders tongue**: Perennial herb to 8.5 cm tall; lf blade 8-35 mm long; spike 5-15 mm long. All regions except NU; all States; New Zealand, New Caledonia, temperate parts of the world.

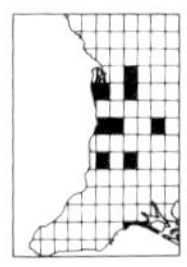

MARSILEACEAE

4. *Marsilea drummondii*, **common nardoo**: Creeping perennial herb; lvs 2-30 cm tall; fruiting body (4a) 4-9 mm long. LE, GT, FR, EA, EP, NL, MU, YP, SE; all mainland States. *M. hirsuta*, short-fruit nardoo, has very shortly stalked fruiting bodies.

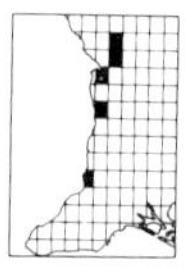

CUPRESSACEAE

5. *Callitris preissii*, **southern cypress-pine**: Broad conical tree to 15 m tall; lvs (5a 3 rings of lvs) to 0.5 mm long; cones (5b) single or grouped, 1.5-4.2 cm wide. LE, GT, FR, EA, EP, NL, MU, YP, KI, SE; WA, NSW, Vic.

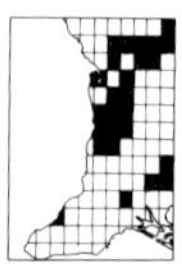

6. *Callitris rhomboidea*, **Oyster Bay pine**: Tree to 5 m tall; lvs (6a: 2 rings of lvs) to 1 mm long; cones (6b) densely clustered, 1.0-2.5 cm wide. KI, SE; all States except WA and the NT.

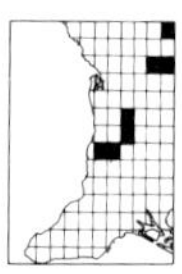

CASUARINACEAE

7. *Allocasuarina muelleriana*, **slaty sheoak**: Tree 0.8-4.0 m tall; branchlet segments 3-9 mm long, with rings of 5-7 teeth (7a); male spike (7b) 1-5 cm long; female cone (7c young, 7d mature) 1-4 cm long, on a stalk 1-8 mm long. Stony siliceous soils. FR, EA, EP, NL, MU, YP, KI, SE; Vic.

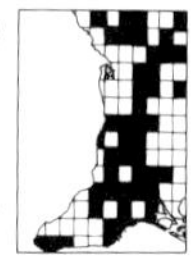

8. *Allocasuarina mackliniana*, a **sheoak**: Shrub 0.6-2.0 m tall; branchlet segments 7-13 mm long with rings of 6-8 teeth (8a); male spike (8b) 1-3 cm long; female cone (8c young, 8d mature) 1.2-2.3 cm long, on a stalk 0-2 mm long. Sandy soils. SE; Vic.

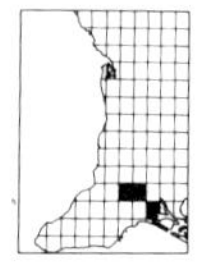

9. *Allocasuarina pusilla*, **dwarf sheoak**: Shrub 0.25-1.0 m tall; branchlet segments 3-9 mm long, with rings of 5-7 teeth (9a); male spike (9b) 0.8-2.0 cm long; female cone (9c young, 9d mature) 1.0-1.5 cm long, stalkless. Note teeth on branches only 0.3-0.5 mm long. Sandy soils. NL, MU, YP, SE; Vic.

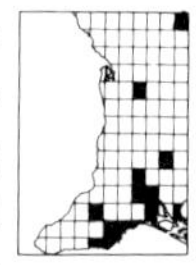

10. *Allocasuarina robusta*, a **sheoak**: Shrub 0.2-1.0 m tall; branchlet segments 7-12 mm long, with rings of 5-7 teeth (10a); male spike (10b) 1.0-4.5 cm long; female cone (10c young, 10d mature) 1.6-2.0 cm long on a stalk 0-2 mm long. Restricted to this area

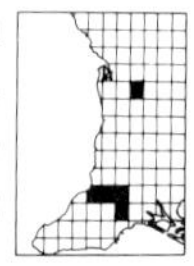

11. *Allocasuarina striata*, **small bull oak**: Shrub or tree 1.5-4.0 m tall; branchlet segments 8-20 mm long, with rings of 5-7 teeth (11a) erect; male spike (11b) 1-3 cm long; female cone (11c young, 11d mature) 1.8-3.3 cm long, on a stalk 3-12 mm long. Note the long cone stalk. Lateritic or sandy soils. KI.

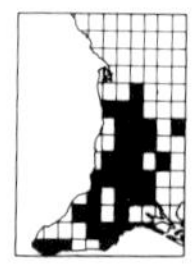

12. *Allocasuarina verticillata*, **drooping sheoak**: Tree 5-9 m tall; branchlet segments 1-4 cm long, with rings of 9-12 teeth (12a); male spike (12b) 4-12 cm long; female cone (12c young, 12d mature) 2.5 cm long. Note the many teeth. Grass woodlands, rocky coastlines. FR, EP, NL, MU, YP, KI, SE; NSW, Vic., Tas.

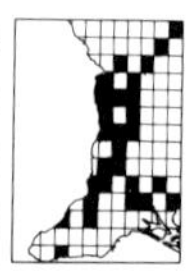

1
1a
2
3
4
4a
5
5a
5b
6
6a
6b
7
7a
7c ♀
7d
b
8
8a
8b ♂
8c ♀
8d
9
9a
9b ♂
9c ♀
9d
10
10a
10b ♂
10c ♀
10d
11
11a
11b ♂
11c ♀
11d
12
12a
12b ♂
12c ♀
12d

URTICACEAE

1. *Parietaria debilis*, **shade pellitory**: Weak annual herb spreading to 30 cm or more; lvs 1.0–3.5 cm long; fls (1a) 2–3 mm long; fruit c. 1.5 mm long. What is treated here as a single species may represent two species but the information to make this split was not available at the time of writing. All regions; all mainland States; America, Africa, Europe and Asia. **P. judaica*, **wall pellitory**, a larger more erect herb may occur here.

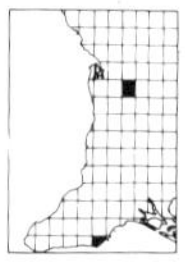

2. *Urtica urens*, **stinging nettle**: Erect annual herb to 70 cm high; lvs 2.5–5.0 cm long, with stinging hairs; fls c. 1 mm long; fruit c. 2.5 mm long. All regions except NW and GT; all States except NT.

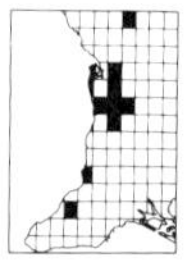

PROTEACEAE

3. *Banksia marginata*, **silver banksia:** Shrub or tree 0.3–10.0 m or more tall; lvs 1.5–6.0 cm long; fl head to 10 cm long, closed fruit 0.4 cm across. EP, NL, MU, KI, SE; all States except WA and NT.

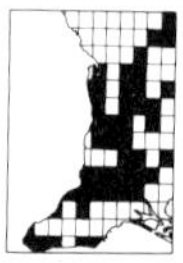

4. *Banksia ornata*, **desert banksia**: Bushy shrub to 3 m tall; lvs 3–11 cm long; fl head to 10 cm long; closed fruit to 2 cm across. Note the toothed lvs. EP, ?MU, KI, SE; Vic.

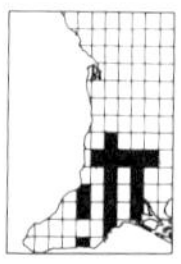

5. *Grevillea lavandulacea*, **lavender grevillea**: Bush 0.3–1.0 m high; lvs 0.2–3.5 cm long; fl 6–11 mm long; fruit (5a) 12–15 mm long excl. the style. FR, NL, MU, SE; Vic. 3 other species, 1 of which is introduced and hybridises with this species.

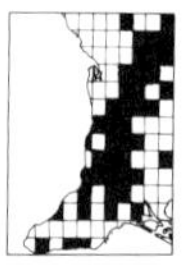

6. *Hakea carinata* a **hakea**: Untidy shrub 1.5–3.0 m high; lvs 5–24 cm long; 'petals' 1.9–2.8 mm long; fruit (6a) 1.3–2.6 cm long incl. the narrow beck. FR, NL, MU, ?KI, SE.

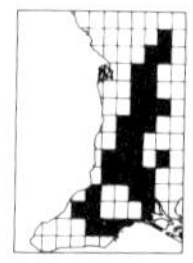

7. *Hakea rostrata*, **beaked hakea:** Shrub 1–5 m high; lvs 2–15 cm long; 'petals' 3.5–5.5 mm long; fruit (7a) 2.2–4.5 cm long. MU, KI, SE; Vic., Tas.

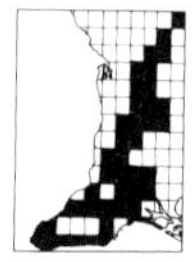

8. *Hakea rugosa*, **dwarf hakea**: Bush or shrub 0.2–2.5 m high; lvs 0.6–6 cm long; 'petals' 2.5–4.2 mm long; fruit (8a) 1.4–2.8 cm long excl. the style. FR, EP, NL, MU, YP, KI, SE; Vic., Tas.

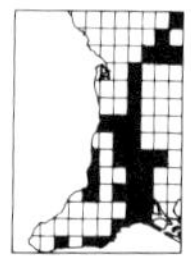

9. **Hakea sericea*, **silky hakea**: Shrub to 4 m high; lvs 1.5–8.0 cm long; fls 4.0–5.5 mm long; fruit (9a) 1.9–2.5 cm long excl. the style. Native to NSW, Vic., Tas.; naturalised in New Zealand, S. Africa and Europe. 3 other species 1 of which is introduced.

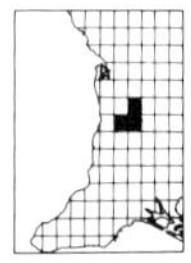

10. *Isopogon ceratophyllus*, **cone-bush**: Prickly shrub 0.2–0.6 rarely to 1.2 m high; lvs rigid, 4–8 cm long; fls 10–20 mm long. Sandy soil. MU, KI, SE; Vic., Tas.

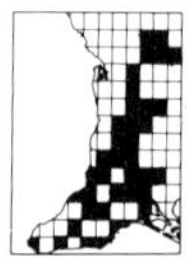

11. *Persoonia juniperina*, **geebung**: Dense shrub to 1 m high; branches white; lvs 1.0–2.8 cm long; fls (11a) 9–13 mm long; fruit fleshy. SE; Vic., Tas.

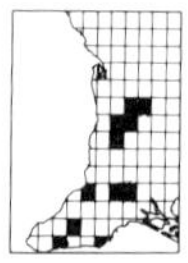

12. *Adenanthos terminalis*, **gland flower**: Prostrate to erect shrub to 2 m high; lvs (12a) 12–21 mm long; fls (12b) 14–21 mm long. Sandy lateritic soil. EP, MU, KI, SE; Vic.

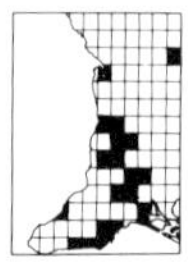

1
1a
2
3
4
5
5a
6
6a
7
7a
8
8a
9
9a
10
11
11a
12
12a
12b

Plate XVIII

SANTALACEAE

1. *Choretrum glomeratum*, a **berry broombush**: Many-stemmed shrub 1-2 m high; lvs 0.5-3.0 mm long; fls (1a) white, 1.0-1.5 mm long; fruit (1b) 3.5-5.0 mm long. FR, EP, MU, YP, KI, SE; WA, NSW, Vic. *C. chrysanthum*, a **berry broombush**, has yellow fls.

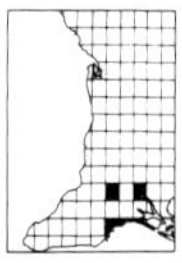

2. *Exocarpos aphyllus*, **leafless ballart**: Shrub or small tree to 3.5 m high; lvs less than 1 mm long; fls in 2-4 mm long spikes; fruit (2a) 7-8 mm diam. All regions except NW and SE; all States except NT.

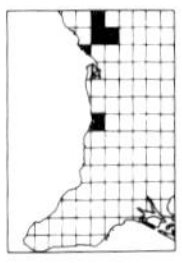

3. *Exocarpos cupressiformis*, **native cherry**: Dense shrub or tree 3-8 m high; lvs (3a) 1-2 mm long; fls in 3-8 mm long spikes; fruit (3b) 4-6.5 mm diam. FR, EP, NL, YP, KI, SE; Qld, NSW, Vic., Tas.

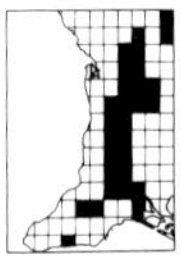

4. *Exocarpos sparteus*, **broom ballart**: Shrub 2-6 m high; branches furrowed; lvs 1.5-7.0 mm long; fls in 4-18 mm long spikes; fruits (4a mature, 4b immature) 3-5 mm diam. NW, FR, EP, NL, MU, YP, SE; all mainland States.

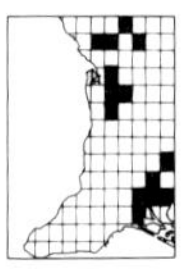

5. *Santalum acuminatum*, **quandong**: Shrub or small tree under 5 m high; lvs 5-19 cm long; fls (5a) 3-4 mm long; fruits (5b) 1.5-3.0 cm diam. All regions; all mainland States.

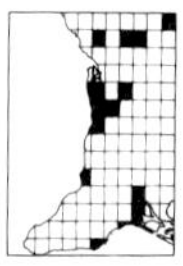

6. *Santalum murrayanum*, **bitter quandong**: Shrub or small tree under 4 m high; lvs 1.5-5.0 cm long; fls 3-4 mm long; fruits (6a) 6.5-2.5 cm diam. FR, EA, EP, MU, YP, SE; WA, NSW.

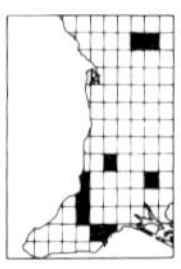

LORANTHACEAE

7. *Amyema miquelii*, **box mistletoe**: Young shoots hairy; lvs 3-25 cm long; fls (7a) 15-28 mm long; fruit (7b) 8-12 mm long. Host plant *Eucalyptus* rarely *Acacia*. All regions except GT and KI; all mainland States.

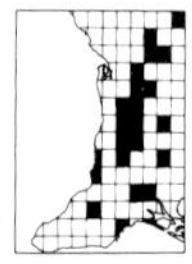

8. *Amyema pendulum*, **drooping mistletoe**: Young shoots hairy; lvs 10-30 cm long; fls (8a) 2-4 mm long; fruit (8b) c. 10 mm long. Host plant *Eucalyptus* rarely *Acacia*. SE; Qld, NSW, Vic. *A. melaleucae*, a mistletoe, has 1 stalkless and 2 stalked fls; host plant *Melaleuca*.

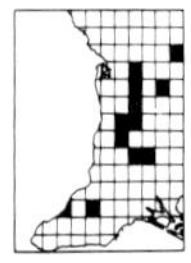

9. *Amyema preissii*, **wire-leaved mistletoe**: Young shoots not hairy, lvs cylindrical, 2-8 cm long; fls (9a) 18-26 mm long; fruit (9b) 8-10 mm long. Host plant usually *Acacia*. All regions except KI; all mainland States.

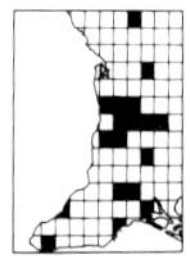

10. *Lysiana exocarpi*, **harlequin mistletoe**: Lvs 3-15 cm long; fls (10a) 25-50 mm long; fruit (10b and 10c) 6-10 mm long. Host plant various but not *Eucalyptus*. All regions; all mainland States.

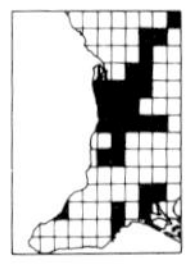

GYROSTEMONACEAE

11. *Gyrostemon australasicus*, **false buck-bush**: Shrubs under 1 m tall, stiff erect branches; lvs 8-30 mm long; fls male (11a) and female (11b) 1-2 mm long; fruit (11c) 4-5 mm long. NU, FR, EP, MU, YP, KI, SE; NSW, Vic.

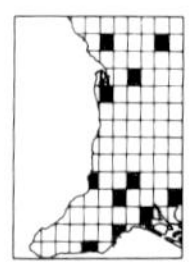

NYCTAGINACEAE

12. *Boerhavia dominii*, a **tar-vine**: Stem to 1 m long; lvs (12a) 10-35 mm long; fls (12b) 1-2 mm long; fruit c. 4 mm long. NW, LE, GT, FR, EA, EP, NL, MU, YP; all mainland States.

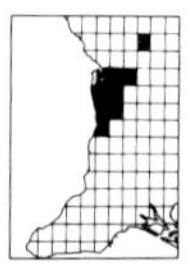

1
1a
1b
2a
2
3
3a
3b
4b
4
5a
5b
5
6a
6
7
7b
7a
8
8a
8b
9a
9
9b
10c
10a
10b
10
11b
♀
♂
11c
11a
11
12a
12b
12

Plate XIX

POLYGONACEAE

1. **Emex australis*, **three-cornered Jack**: Low flat growing plant with stems to 50 cm long; lvs to 6 cm long; fls male (1a) and female (1b) 2-3 mm diam.; 'fruit' (1c) 7-10 mm diam. LE, GT, FR, EA, EP, NL, MU, ?YP, KI, SE; all mainland States; native to S. Africa; naturalised in N. America, Hawaii, New Zealand.

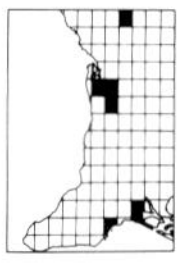

2. *Muehlenbeckia adpressa*, **climbing lignum**: Twining plant; lvs 1.5-6.0 cm long; fls (2a) c. 2 mm diam.; fruit 3-4 mm long, shiny, black and smooth. LE, GT, EP, NL, MU, YP, KI, SE; WA, NSW, Vic., Tas.

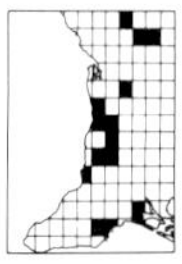

3. *Muehlenbeckia cunninghamii*, **lignum**: Shrub 1-2 m tall; lvs c. 1 mm long; fls male (2a) and female (3b) 2-3 mm long; fruit 3-4 mm long, shiny, smooth, brown-black. All regions except NU and KI; all mainland States.

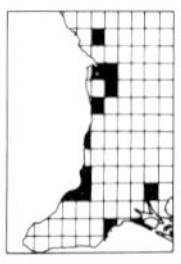

4. *Muehlenbeckia gunnii*, **coastal lignum**: Climbing plant; lvs 3-8 cm long; fls (4a) 2-3 mm long; fruit c. 5 mm long, matt, black. EP, MU, YP, KI, SE; Tas.

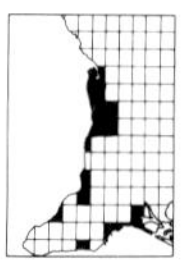

5. **Polygonum aviculare*, **wireweed**: Flat plant; stems to 1 m long; lvs 1-4 cm long; stipules (5a); fls (5b) 1-2 mm long; fruit to 2.5 mm long. LE, GT, FR, EA, EP, NL, MU, YP, SE; all mainland States; native to Europe; naturalised widely including New Zealand.

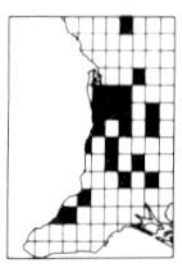

6. *Polygonum lapathifolium*, **pale knotweed**: Plant to 1 m high; lvs 7-18 cm long; stipules (6a); fls 2-7 cm long; fruit to 2 mm long. LE, EA, MU, SE; all States except WA; New Zealand, Africa.

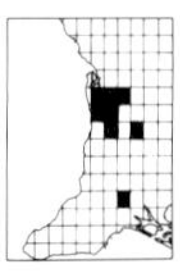

7. **Polygonum salicifolium*, **slender knotweed**: Flat or erect plant; stems to 1 m long; lvs 5-10 cm long; stipules (7a); fls 2-3 mm long; fruit 2.0-2.5 mm long. LE, FR, NL, MU, SE; all mainland States; possibly native to Europe and Asia.

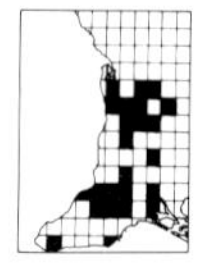

8. **Rumex acetosella*, **sorrel**: Slender plant to 60 cm tall; lvs 2-10 cm long; fls male (8a) and female (8b) c. 2 mm diam.; 'fruit' 1.0-1.5 mm long. EP, NL, MU, KI, SE; all States except NT; native to Mediterranean; widely naturalised in temperate regions.

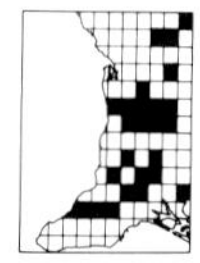

9. **Rumex conglomeratus*, **clustered dock**: Plant to 1 m tall; lower lvs 3-13 cm long; fls 1-2 mm long; 'fruit' (9a) 2-3 mm long. EP, NL, SE; all States except NT; almost cosmopolitan.

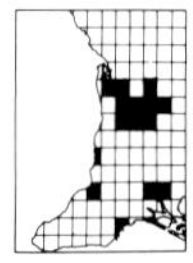

10. **Rumex crispus*, **curled dock**: Stems 30-150 cm tall; lower lvs 3-30 cm long; fls 1-2 mm long; 'fruit' (10a) 3-6 mm long. ?EA, EP, NL, MU, YP, KI, SE; all States; New Zealand; native to Europe and N. Africa.

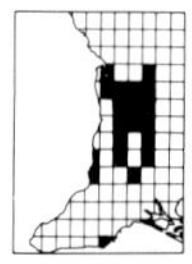

11. **Rumex obtusifolius*, **broad-leaved dock**: Stems to 1 m tall; lvs 1-3 cm long, partly hairy; fls 1-2 mm long; 'fruit' (11a) 4-6 mm long. EP, YP, SE; all States except NT; native to Europe and Asia; naturalised in New Zealand.

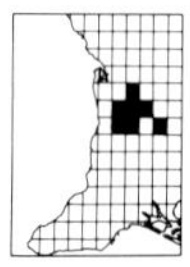

12. **Rumex pulcher*, **fiddle dock**: Stems to 60 cm tall; lvs 3-13 cm long; fls 1-2 mm long; 'fruit' (12a) 3-5 mm long. FR, EP, NL, SE; all States except NT and Qld; native to Europe and Asia. 5 other species.

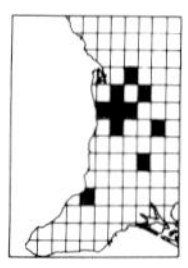

1a ♂
1c
1b
1
2a
2
3b
♂
3a
♀
3
4a
4a
4
5a
5b
5
6
6a
7a
7
8b ♀
8c
8a ♂
8
9a
9
10a
10
11a
11
12a
12

Plate XX

AIZOACEAE

1. *Carpobrotus rossii*, **karkalla**: Stems to 1 m long; lvs 3.5–10.0 cm long; fls 3.5–5.5 cm diam.; fruit 1.6–2.2 cm long. NU, FR, EP, NL, MU, YP, KI, SE; WA, Vic., Tas. 3 other species.

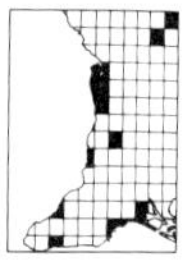

2. *Disphyma crassifolium*, **round-leaved pig-face**: Stems to 1 m long; lvs 2–3 cm long; fls 2–5 cm diam.; fruit (2a) 5–12 mm diam. Along coast and saline areas. All regions except NW and FR; all States except NT

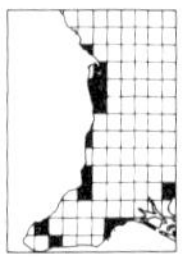

3. **Galenia pubescens*, **coastal galenia**: Stems to 1 m long, with appressed hairs; lvs 3–20 mm long; fls (3a) 2–3 mm long; fruit 1–2 mm long. Especially near the sea. LE, NU, FR, EP, NL, YP; WA, Vic.; native to S. Africa.

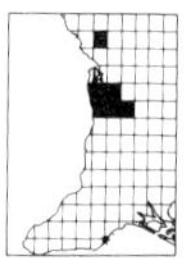

4. **Galenia secunda*, a **galenia**: Stems 20–50 cm long, with spreading hairs; lvs 3–12 mm long; fls (4a) 2–3 mm long; fruit 1–2 mm long. EP, NL, MU, YP, SE; Vic., Tas.; native to S. Africa.

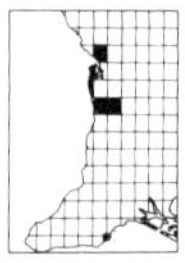

5. *Tetragonia implexicoma*, **bower spinach**: Prostrate undershrub or climber on shrubs; lvs 2–4 cm long; fls (5a) 5–20 mm long; fruits (5b) to 5 mm diam. NU, EP, NL, MU, YP, KI, SE; WA, Vic., Tas. **T. decumbens*, sea spinach, has its fls in clusters of 3–5 and 4-winged fruit.

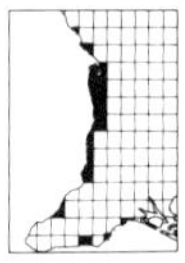

6. **Mesembryanthemum crystallinum*, **common iceplant**: Prostrate; basal lvs 6–12 cm long; stem lvs 2–4 cm long; fls c. 10 mm long; fruits 8 mm long. NU, GT, EP, NL, MU, YP, KI, SE; all States except NT and Qld; native to S. Africa.

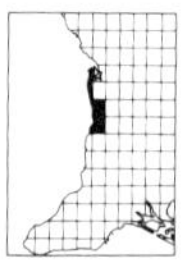

PORTULACACEAE

7. *Calandrinia eremaea*, **small purslane**: Stems to 20 cm long; lvs 1.0–5.5 cm long; fls (7a) 3–7 mm long; fruit 3–5 mm long. All regions except KI; all mainland States.

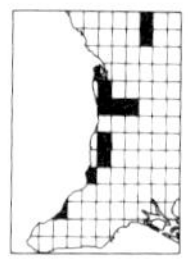

8. *Calandrinia granulifera*, **pigmy purslane**: Stems to 5 cm long; lvs 4–12 mm long; fl (8a) 2.0–3.5 mm long; fruit 1.5–2.2 mm long. EP, MU, SE; WA, NSW, Vic. 4 other species.

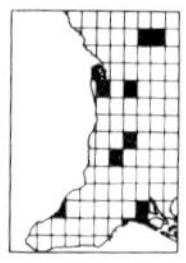

9. *Portulaca oleracea*, **munyeroo**: Stems to 25 cm long; lvs 1.0–2.5 cm long; fls 4–7 mm long; fruit (9a and 9b) 3–5 mm long. NW, LE, GT, FR, EA, YP, SE; all States, temperate and tropical regions of the world.

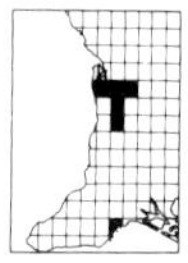

CARYOPHYLLACEAE

10. **Petrorhagia velutina*, **velvet pink**: Plant to 60 cm tall; lvs c. 3 cm long; fls (10a) c. 12 mm long; fruit 5–7 mm long. FR, EP, MU, YP, KI, SE; all States except NT; native to southern Europe. **P. nanteuilii*, **proliferous pink**, mainly lacks the glandular hairs on the stems.

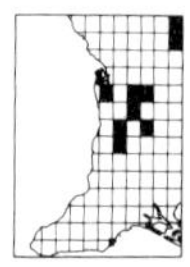

11. **Silene gallica*, **French catchfly**: Stems to 75 cm tall; lvs c. 5 cm long; fls c. 5 mm long; fruit 6–9 mm long. FR, EP, NL, MU, YP, KI, SE; all States except NT; native to Europe.

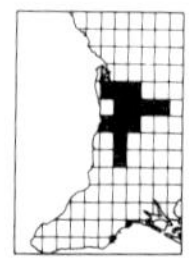

12. **Silene vulgaris*, **bladder campion**: Stems to 70 cm high; lvs c. 8 cm long; fls c. 1 cm long; fruit c. 1.4 cm long. EP, MU, YP, KI, SE; all States except NT, Qld; New Zealand; native to Europe, the Mediterranean and western Asia. Up to 6 other species.

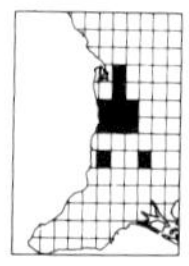

(*Continued*)

1
2
2a
3
3a
4
4a
5
5a
5b
6
7
7a
8
8a
9
9a
9b
10
10a
11
12

Plate XXI

CARYOPHYLLACEAE
(*Continued*)

1. **Cerastium glomeratum*, **broad-leaved mouse-ear chickweed**: Annual herb to c. 45 cm tall; lvs 5-25 mm long; fl (1a) sepals c. 5 mm long. FR, EP, NL, MU, YP, KI, SE, all States; native to Europe.

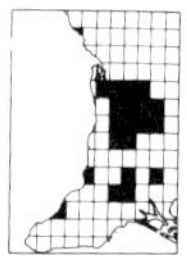

2. **Cerastium semidecandrum*, **small mouse-ear chickweed**: Annual herb to c. 20 cm tall; lvs c. 18 mm long; fl (2a) sepals c. 3.5 mm long; fruit (2b) 4-7 mm long. EP, NL, MU, YP, KI, SE; NSW, Vic.; native to Europe and Middle East.

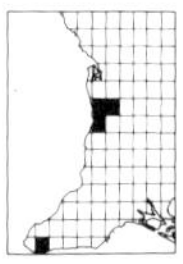

3. **Moenchia erecta*, **erect chickweed**: Annual herb 3-20 cm tall; lvs 5-15 mm long; fl (3a) sepals 5-6 mm long. MU, SE; NSW, Vic., Tas.; native to Europe.

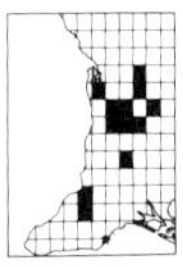

4. **Sagina apetala*, **common pearlwort**: Annual herb to c. 10 cm tall; lvs to 10 mm long; lf base (4a) with hairs; fls (4b) 1.5-2.5 mm long. GT, FR, EP, NL, MU, YP, KI, SE; all mainland States except NT; native to Europe.

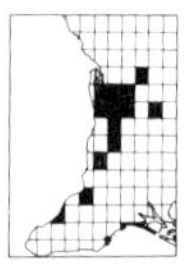

5. **Sagina maritima*, **sea pearlwort**: Annual herb to c. 10 cm tall; lvs to 8 mm long; lf base (5a) without hairs; fls (5b) 2-3 mm long. ?NU, EP, MU, YP, KI, SE; WA, ?NSW, Vic., Tas.; native to Europe. **S. procumbens*, **spreading pearlwort**, is a spreading plant on the ground.

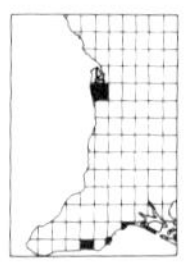

6. *Scleranthus pungens*, **prickly knawel**: Perennial herb to c. 30 cm tall; lvs c. 15 mm long; fls (6a) 2-4 mm long; stamens 5. FR, EA, EP, NL, MU, YP, KI; NSW. *S. miniscculus*, **cushion knawel**, has stalked fls and 2 stamens.

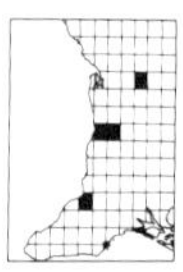

7. **Polycarpon tetraphyllum*, **allseed**: Annual herb to c. 23 cm tall; lvs 2.5 cm long; fl (7a) sepals c. 2 mm long. FR, EP, MU, YP, KI, SE; all States except NT; native to Europe.

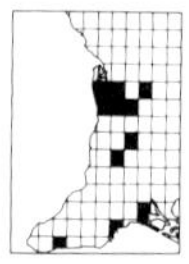

8. **Spergularia marina*, **salt sand-spurrey**: Herb 3-35 cm tall; lvs 1-4 cm long; fls (8a) 6-10 mm wide, petals pink to white; fruits (8b) 3-6 mm long; seeds (8c) wingless. LE, GT, FR, EA, EP, NL, MU, YP, KI, SE; NSW, Vic., Tas.; native to Europe and temperate parts of the Northern Hemisphere.

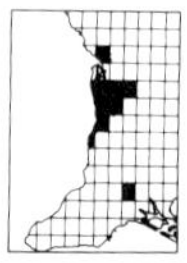

9. **Spergularia media*, **coast sand-spurrey**: Herb 5-40 cm tall; lvs 1-7 cm long; fls (9a) 8-14 mm diam., petals white, pink or violet; seeds (9b) winged. EP, NL, MU, YP, KI, SE; Vic., Tas.; native to Europe and Asia. 2 other species.

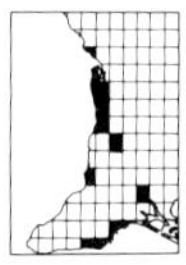

10. **Spergula arvensis*, **corn spurrey**: Annual 20-40 cm tall; lvs 1-3 cm long; fls (10a) 3-5 mm long, on c. 2.5 cm long stalks. SE; all States except NT; native to Europe.

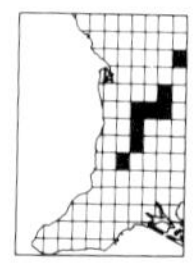

11. **Stellaria media*, **chickweed**: Annual c. 40 cm tall; lvs c. 18 mm long; fls (11a) 8-10 mm diam. FR, EP, NL, MU, YP, KI, SE; all States; native to Europe.

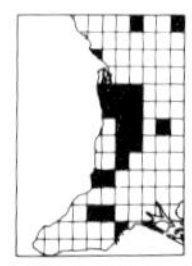

12. **Stellaria palustris*, **swamp starwort**: Herb c. 60 cm tall; lvs 0.3-4.0 cm long; fls (12a) c. 10 mm long or longer. NL, MU, KI, SE; all States except WA and NT; native to Europe and Asia. 2 or 3 other species.

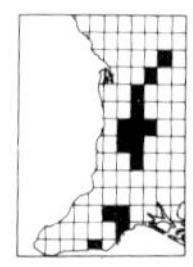

1
1a
2
2a
2b
3
3a
4
4a
4b
5
5a
5b
6
6a
7
7a
8
8a
8b
8c
9
9a
9b
10
10a
11
11a
12
12a

Plate XXII

CHENOPODIACEAE

1. *Atriplex cinerea*, **grey saltbush**: Shrub to 1.5 m tall; lvs 2.5-4 cm long; 'fruit' (1a and 1b) 6-10 mm long. NU, EP, NL, YP, KI, SE; all States except NT and Qld.

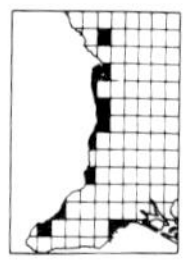

2. *Atriplex paludosa*, **marsh saltbush**: Shrub to 1 m tall; lvs 1-4 cm long; 'fruit' (2a) 7-12 mm long. NU, EP, MU, YP, KI, SE; WA, Vic., Tas.

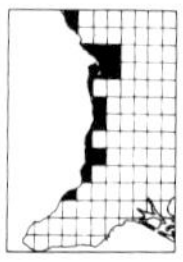

3. **Atriplex prostrata*, **hastate orache**: Annual herb to 50 cm tall; lvs 1.5-5.0 cm long; 'fruit' (3a) 3-6 mm long; seed (3b) c. 1.5 cm diam. GT, EP, NL, MU, YP, SE; all States except NT and Qld; native to Europe. *A. australasica*, native orache has narrower lvs.

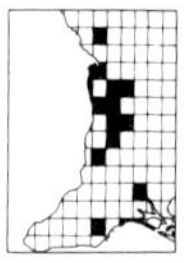

4. *Atriplex semibaccata*, **berry saltbush**: Prostrate herb with woody taproot; lvs 1-2 cm long; 'fruit' (4a, 4b and 4c) 2-5 mm long. FR, EA, EP, NL, MU, YP, KI, SE; all States. *A. suberecta*, lagoon saltbush, has coarsely toothed lvs.

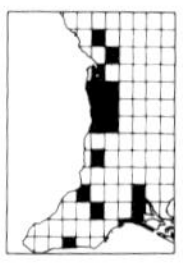

5. **Chenopodium album*, **fat hen**: Annual herb to 1 m tall; lvs 2-4 cm long; fruit (5a) c. 2-3 mm diam.; seed c. 1.2 mm diam. LE, FR, EP, NL, YP, KI, SE; all States; cosmopolitan.

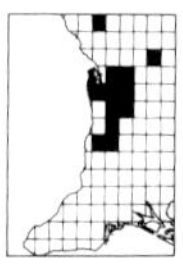

6. *Chenopodium desertorum*, a **goosefoot**: Prostrate or erect herb to 50 cm tall; lvs 5-20 mm long; seed c. 1.5 mm diam. All regions; all mainland States.

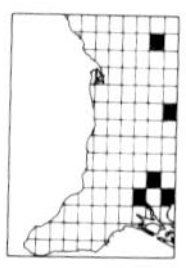

7. **Chenopodium glaucum*, **glaucous goosefoot**: Prostrate or erect annual to 0.5 m tall; lvs 1-3 mm long; seed (7a) c. 1 mm diam. MU, YP, KI, SE; all States except NT and Qld; cosmopolitan.

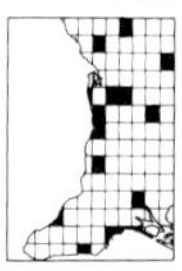

8. **Chenopodium multifidum*, **scented goosefoot**: Prostrate to erect perennial herb to 1 m tall; lvs (8a) to 6 cm long; fruit (8b) 2.5 mm long. EP, MU, YP, SL; all States except NT and Tas; native to S. America.

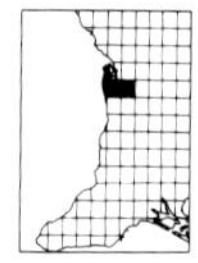

9. **Chenopodium murale*, **nettle-leaved goosefoot**: Branched annual to 1 m high; lvs (9a) 2-8 cm long; seed 1.0-1.5 mm diam. All regions; all States, cosmopolitan.

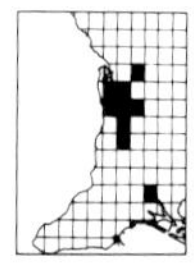

10. *Chenopodium pumilio*, **clammy goosefoot**: Prostrate or erect plant to 30 cm tall; lvs 1-2 cm long; fruit (10a) 1 mm long; seed c. 0.5 mm long. All regions except NU; all States. 4 other species.

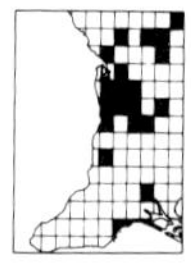

11. *Rhagodia candolleana*, **seaberry saltbush**: Weak spreading shrub to 1.0-1.5 m tall; lvs (11a) 10-25 mm long; fls c. 1 mm diam.; fruit (11b) 3-4 mm diam.; seed 1.5-2.5 mm diam. NW, NU, EP, NL, MU, YP, KI, SE; all States except NT and Qld. *R. crassifolia*, fleshy saltbush, has curved fleshy lvs.

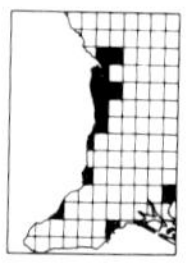

12. *Rhagodia parabolica*, **fragrant saltbush**: Rounded shrub 1-2 m tall; lvs (12a) 1.5-4.0 cm long; fls c. 1 mm diam.; fruit (12b) 2-3 mm diam.; seed c. 1.8 mm diam. All regions except LE and KI; all States except WA and Tas. 2 other species.

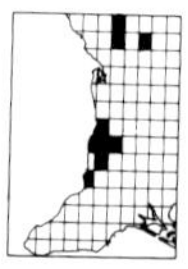

(*Continued*)

1
1a
1b
2a
2
3a
3b
3
4b
4c
4
4a
5
5a
6
7
7a
8a
8
8b
9a
9
10
10a
11b
11
11a
12b
12a
12

Plate XXIII

CHENOPODIACEAE

(*Continued*)

1. **Beta vulgaris*, **beet**: Prostrate to erect herb with slender taproot; lvs to 5 cm long; fl bud (1a) c. 1 mm long; fruit (1b) c. 7 mm diam. in clusters. YP, KI; NSW, Vic.; native to Europe.

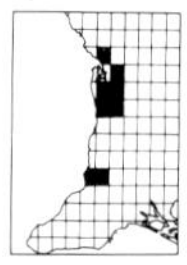

2. *Einadia nutans*, **climbing saltbush**: Scrambling perennial with woolly base; lvs (2a) usually 1-2 cm long; infl 2-4 cm long; fruit (2b) c. 4 mm diam. All regions except NU; all States.

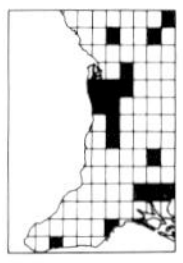

3. *Maireana aphylla*, **cotton bush**: Shrub 0.5-2.0 m tall; lvs 1-4 mm long; 'fruits' (3a) 1-2 mm long, c. 8 mm diam. NW, LE, GT, FR, EA, EP, NL, MU, YP; all mainland States.

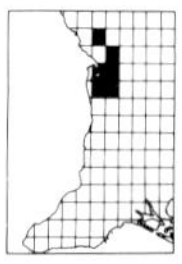

4. *Maireana brevifolia*, **small-leaved bluebush**: Shrub to 1 m tall; lvs 2-5 mm long; 'fruit' (4a) 4-5 mm diam. NU, FR, EA, EP, NL, MU, YP, SE; all mainland States.

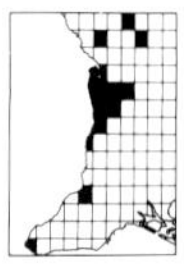

5. *Maireana enchylaenoides*, **wingless fissure weed**: Shrub to 20 cm tall; lvs 4-10 mm long; 'fruit' (5a) c. 4 mm diam. FR, EP, NL, MU, YP, KI, SE; all States except NT and Tas.

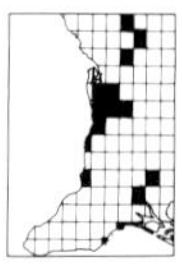

6. *Maireana oppositifolia*, **heathy bluebush**: Compact shrub to 1 m tall; lvs to 4 mm long; 'fruit' (6a) c. 7 mm diam. NW, NU, GT, EA, EP, NL, MU, YP, KI, SE; WA.

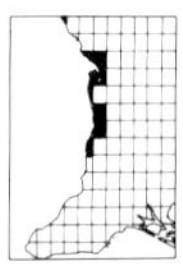

7. *Sclerolaena diacantha*, **grey copperburr**: Shrub to 30 cm tall; lvs fleshy, 10-15 mm long; 'fruit' (7a) c. 5 mm long (incl. spines). All regions except KI and SE; all mainland States.

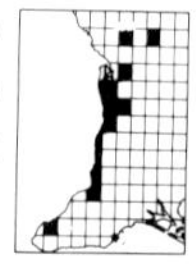

8. *Halosarcia flabelliformis*, a **samphire**: Woody perennial to 20 cm tall; infl (8a) terminal, 2-4 cm long, segments barrel-shaped, c. 5 mm long; seed (8b) c. 1.8 mm long. EP, NL, YP; Vic.

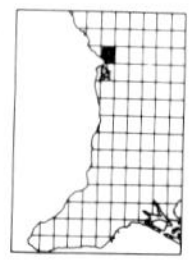

9. *Halosarcia halocnemoides*, **grey samphire**: Spreading or erect shrub to 0.5 m tall; infl terminal, to 2 cm long, segments 3-5 mm long; seed (9a) c. 1 mm long. NW, LE, NU, GT, EA, NL, YP, SE; all States except NSW and Tas.

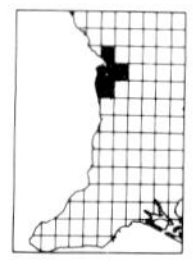

10. *Halosarcia pergranulata*, a **samphire**: Spreading shrub to 50 cm tall; fls (10a) infl terminal, to 5 cm long, segments 5-10 mm long; seed (10b) c. 1 mm diam. All regions except NW and FR; all mainland States.

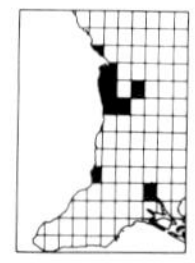

11. *Sarcocornia quinqueflora*, **beaded glasswort**: Prostrate to erect perennial to 0.5 m tall; infl terminal, 2-5 cm long, segments 5-15 mm long; seeds (11a) c. 1 mm long. EP, NL, MU, YP, KI, SE; all States except NT.

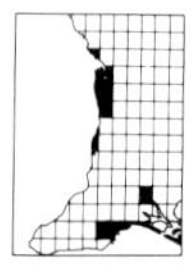

12. *Sclerostegia arbuscula*, **shrubby glasswort**: Branched shrub to 2 m tall; infl 1-3 cm long, segments (12a) c. 5 mm long. EP, MU, YP, KI, SE; all States except NT and Qld.

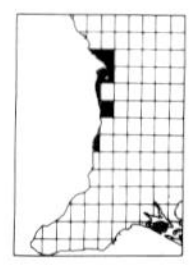

(*Continued*)

1b
1a
1
2a
2b
2
3a
3
4a
4
5
5a
6a
6
7a
7
8
8b
8a
9
9a
10b
10
10a
11
11a
12
12a

Plate XXIV

CHENOPODIACEAE
(Continued)

1. *Enchylaena tomentosa*, **ruby saltbush**: Shrub to c. 1 m tall; lvs 7–20 mm long; fls c. 1 mm long; 'fruit' green, red (1a) or yellow (1b), c. 5 mm diam. All regions; all mainland States.

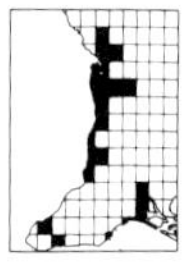

2. *Salsola kali*, **buckbush**: Prickly annual bush c. 60 cm tall; lvs 1–3 cm long; fls 3–4 mm long; 'fruit' (2a) 3.5–10.0 mm diam. incl. the wings. All regions; all mainland States; Europe and Asia.

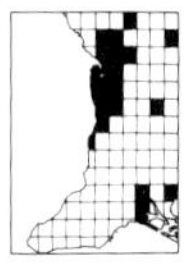

3. *Suaeda australis*, **austral seablite**: Perennial shrub to 1 m tall; lvs c. 1–3 cm long; fls male (3a), female (3b) c. 1 mm long; 'fruit' c. 3 mm diam. Grows along the coast. EP, NL, MU, YP, KI, SE; all States except NT. **S. baccifera* has translucent lf margins.

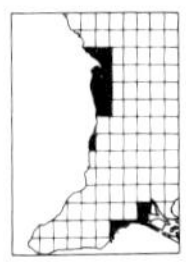

4. *Threlkeldia diffusa*, **coast bonefruit**: Perennial shrub to 30 cm tall; lvs c. 5–30 mm long; fls (4a) c. 2 mm long; 'fruit' (4b) 1.5–2.5 mm long. Usually grows along the coast. NU, EP, NL, MU, YP, KI, SE; WA, Vic., Tas.

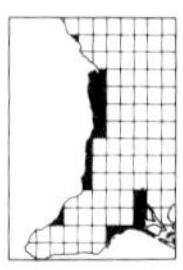

AMARANTHACEAE

5. *Alternanthera denticulata*, **lesser joyweed**: Annual often prostrate herb; lvs 2–5 cm long; fls 2–3 mm long; fruit (5a with bracts and 'petals') c. 1.5–2.5 mm long. NW, LE, GT, FR, EA, NL, MU, KI, SE; all States. **A. pungens*, **khaki weed**, has prickly lvs.

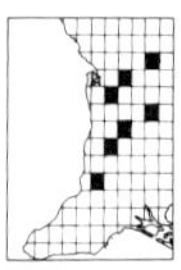

6. **Amaranthus albus*, **tumbleweed**: Annual herb 20–80 cm tall; lvs 5–20 mm long; fls c. 2 mm long; fruit (6a) c. 1.5–2.0 mm long. NL, MU, SE; WA, NSW, Vic.; native to N. America.

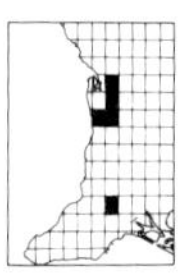

7. **Amaranthus muricatus*, **rough-fruited amaranth**: Perennial herb often 10–60 cm tall; lvs 1–9 cm long incl. the stalk; fls (7a) c. 2 mm long; fruit c. 3.0–3.5 mm long. EP, NL, MU, YP; NSW, Vic.; native to S. America.

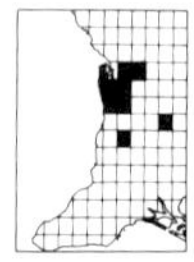

8. **Amaranthus retroflexus*, **red-root amaranth**: Annual herb to c. 1 m tall; lvs 3–10 cm long incl. the stalk; fls 2–3 mm long; fruit (8a with 'petals') c. 2–3 mm long. EP, NL, YP, SE; all States except NT and Qld; native to N. America. 5 other species.

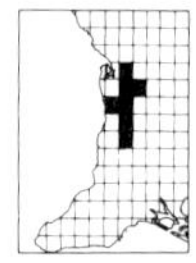

9. *Ptilotus erubescens*, **hairy heads**: Perennial herb to 40 cm tall; lvs to 12 cm long; spikes 1.5–4.0 cm long; fls to c. 2 cm long. NL, MU; NSW, Vic.

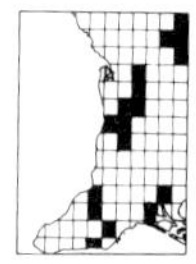

10. *Ptilotus nobilis*, **yellow tails**: Perennial herb 12–40 cm long; lvs 3–7 cm long; spikes 4–7 cm long; fls 2.0–2.5 cm long. NW, LE, NU, GT, FR, EA, EP, NL, MU; all mainland States.

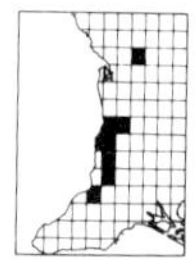

11. *Ptilotus polystachyus*, **long tails**: Annual or perennial herb to c. 1.5 m tall; lvs to 21 cm long; spikes greenish or reddish, 3–15 cm long; fls to 2 cm long. NW, LE, NU, GT, FR, EA, EP, NL, MU; all mainland States.

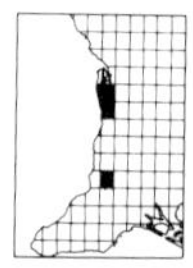

12. *Ptilotus spathulatus*, **pussytails**: Perennial herbs with stems to c. 40 cm long along the ground; lvs to 10 cm or more long; spike 2–9 cm long; fls 7.5–14.0 mm long. FR, EP, NL, MU, YP, KI, SE; all States except NT and Qld.

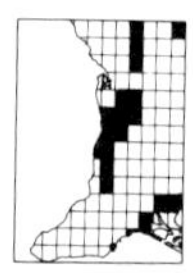

1a
1b
1
2a
2
3b
♀
3
3a ♂
4a
4b
4
5a
5
6a
6
7a
7
8
8a
9
10
11
12

LAURACEAE

1. *Cassytha glabella*, **slender dodder-laurel**: Rootless twining perennial; lvs scale-like; fls white, c. 1.5 mm long; fruit elongate, c. 10 mm long. EP, NL, MU, YP, KI, SE; all States except NT and Qld.

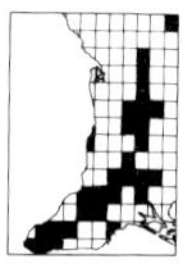

2. *Cassytha melantha*, **strangle vine**: Rootless twining perennial; lvs scale-like; fls covered by blackish hairs, 2-3 mm long; fruit (2a) globular, 10-15 mm diam. NU, GT, EP, NL, MU, YP, KI, SE; all States except NT and Qld.

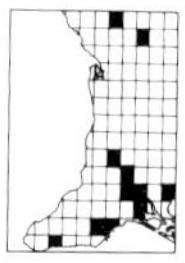

3. *Cassytha pubescens*, **downy dodder-laurel**: Rootless twining perennial; lvs scale-like; fls with mostly grey hairs, 1-2 mm long; fruit (3a) globular, 6-8 mm long. EP, NL, MU, KI, SE; all States except WA and NT; ?New Zealand.

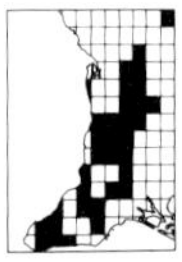

RANUNCULACEAE

4. *Clematis microphylla*, **old man's beard**: Climbing perennial; lvs with 3 or more leaflets 0.8-3.0 cm long; fls with creamy-white 'petals' 1.5-2.5 cm long; fruit (4a) with feathery styles 2-4 cm long. FR, EA, EP, NL, MU, YP, KI, SE; all States except NT. **C. vitalba*, **travellers joy**, has smaller fls.

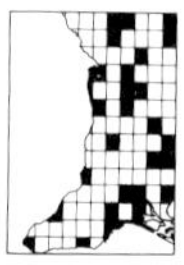

5. *Ranunculus lappaceus*, **common buttercup**: Tufted perennial herb 4-170 cm tall; lvs 1.5-8.0 cm long excl. the stalk; fls with petals 7-17 mm long; fruit (5a) somewhat flattened, 1.7-3.5 mm long. FR, ?EP, NL, SE; all States except WA and NT; ?New Zealand.

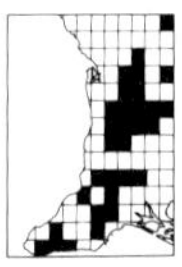

6. **Ranunculus muricatus*, **sharp buttercup**: Annual herb 10-50 cm tall; lvs (6a) 1.5-5.0 cm long excl. the stalk; fls (6b) with petals 7-9 mm long; fruit (6c) spiny, flattened, 7-8 mm long. FR, EP, NL, MU, SE; all States except NT; native to Europe.

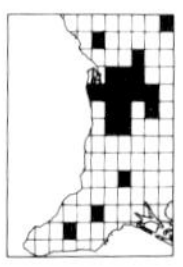

7. *Ranunculus pachycarpus*, **thick-fruited buttercup**: Tufted perennial herb 6-40 cm tall; lvs (7a) 8-45 mm long excl. the stalk; fls (7b) with petals 6-15 mm long; fruit (7c) globular, 2.5-5.0 mm long. FR, NL, MU, SE; NSW, Vic.

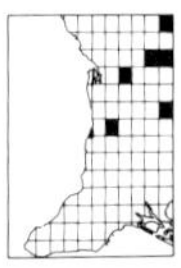

8. **Ranunculus repens*, **creeping buttercup**: Perennial herb with long stolons, 15-60 cm tall; lvs 3-9 cm long excl. the stalk; fls (8a) with petals 6-12 mm long; fruit (8b head of fruits) compressed, 2.5-3.5 mm long. NL; NSW, Vic., Tas.; native to Europe and Asia.

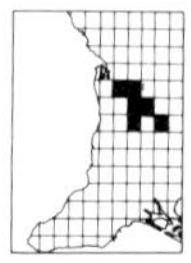

9. *Ranunculus amphitrichus*, **river-buttercup**: Perennial herb with long stolons, to 40 cm tall; lvs 5-30 mm long excl. the stalk; fls with petals 3-4 mm long; fruit (9a) somewhat flattened, c. 2 mm long. In wet places. FR, NL, MU, KI, SE; NSW, Vic., Tas.; New Zealand.

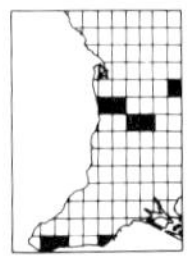

10. *Ranunculus sessiliflorus*, **small-flowered buttercup**: Annual herb 2-35 cm tall; lvs 3-20 mm long excl. the stalk; fls 2-3 mm diam., with 0-2 petals; fruit (10a head of fruits) compressed, 1.5-2.0 mm long; note that the fls are almost stalkless. NW, LE, FR, EP, NL, MU, YP, KI, SE; all States except NT.

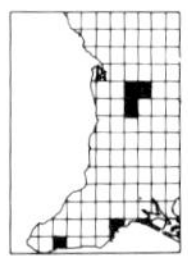

11. **Ranunculus trilobus*, **large annual buttercup**: Annual herb 20-80 cm tall; lvs (11a) usually 2-4 cm long excl. the stalk; fls (11b) with petals 4-5 mm long; fruit (11c) compressed, c. 2-3 mm long. EP, MU, SE; all States except NT and Qld; native to Europe. 6 other species.

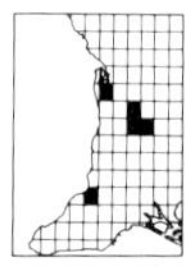

RESEDACEAE

12. **Reseda lutea*, **cut-leaved mignonette**: Sprawling or erect perennial herb to 80 cm tall; lvs (12a) to 10 cm long incl. the stalk; fls (12b) with 6 petals and sepals, the largest petal (12c) deeply 3-lobed, c. 3 mm long; fruit (12d) 5-13 mm long. EP, NL, MU, YP, SE; all States except NT; native to the Mediterranean and Asia. 3 other species.

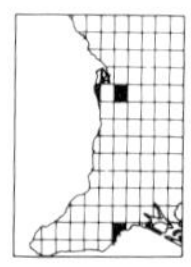

1
2a
2
3
3a
4a
5
5a
6
6a
6b
6c
7
7a
7b
7c
8
8a
8b
9
9a
10
10a
11
11a
11b
11c
12
12a
12b
12c
12d

Plate XXVI

DROSERACEAE

1. *Drosera auriculata*, **tall sundew**: Perennial herb 15–70 cm tall; lvs 5–16 mm long incl. the stalk; calyx (1a) lacking hairs; fls (1b) white or pink, with petals 5–8 mm long. FR, EP, NL, MU, YP, KI, SE; all States except WA and NT; New Zealand. *D. peltata*, **pale sundew**, has a hairy calyx (1c).

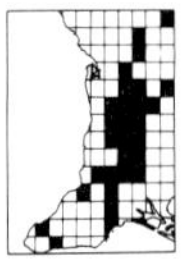

2. *Drosera binata*, **forked sundew**: Annual herb to 50 cm tall; lvs 6–40 cm long incl. the stalk; fls (2a) white, with petals 9–14 mm long. KI, SE; all States except WA and NT; New Zealand.

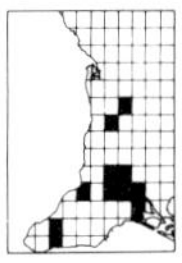

3. *Drosera glanduligera*, **scarlet sundew**: Annual herb 3–5 cm tall; lvs 12–20 mm long incl. the stalk; fls (3a) orange to red, with petals 4–5 mm long. FR, EP, NL, MU, YP, KI, SE; all States except NT and Qld.

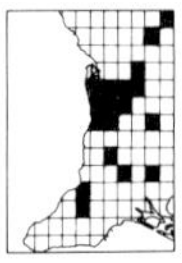

4. *Drosera macrantha*, **climbing sundew**: Perennial climbing herb to 60 cm tall; lvs 1.5–6.0 cm long incl. the stalk; fls (4a and 4b) white to red, with petals 6–14 mm long. FR, EP, NL, MU, YP, KI, SE; Vic., Tas.

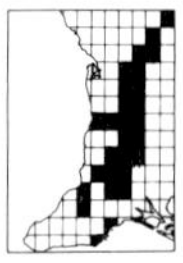

5. *Drosera pygmaea*, **tiny sundew**: Annual herb 5–45 mm tall; lvs 3–7 mm long incl. the stalk; fls white, with petals 2.0–2.5 mm long. EP, YP, KI, SE; all States except NT.

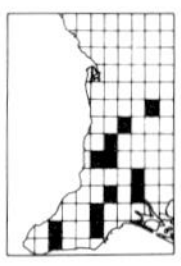

6. *Drosera whittakeri*, **scented sundew**: Perennial herb 2–5 cm tall; lvs (6a) 15–35 mm long, green or red; fls (6b and 6c) white, with petals 10–12 mm long. NL, MU, KI, SE; Vic.

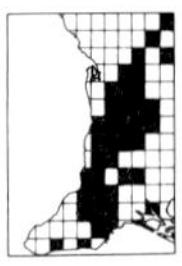

PAPAVERACEAE

7. **Glaucium flavum*, a **horned-poppy**: Annual robust herb to 50 cm tall; lvs to 16 cm long; fls (7a) yellow, with petals 3–4 cm long; fruit (7b) 8–25 cm long. YP; NSW, Vic.; native to the Middle East. **G. corniculatum*, **red horned-poppy**, has red fls.

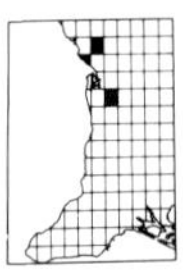

8. **Papaver dubium*, **long-headed poppy**: Annual herb 20–60 cm tall; lvs to 20 cm long incl. the stalk; fls (8a) reddish, with petals 15–30 mm long; fruit (8b) 1.5–2.0 cm long. NL, YP, SE; NSW, Vic., Tas.; native to Europe, Africa and Asia.

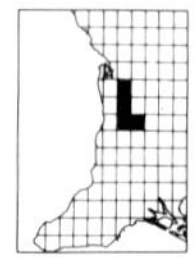

9. **Papaver hybridum*, **rough poppy**: Annual herb to 30 cm tall; lvs to 20 cm long incl. the stalk; fls (9a) reddish with a dark centre, with petals 1–2 cm long; fruit (9b) 1.0–1.25 cm long. LE, NU, FR, EA, EP, MU, YP; all States; native to Europe, Africa and Asia. 3 other species.

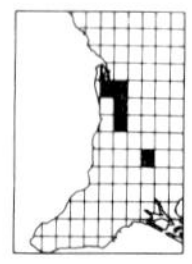

FUMARIACEAE

10. **Fumaria bastardii*, **Bastards fumitory**: Scrambling annual herb with stems to 80 cm long; lvs to 20 cm long incl. the stalk; fls (10a) pink to deep-pink (usually darker than *F. muralis*), 9–12 mm long, more open at the mouth than *F. muralis*; infl with 15–25 fls. EP, MU, YP, KI; NSW, Vic., Tas.; native to Europe and the Mediterranean.

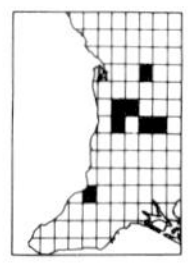

11. **Fumaria capreolata*, **white fumitory**: Scrambling annual herb with stems 30–120 cm long; lvs to 25 cm long incl. the stalk; infl with 9–20 fls; fls (11a) white except the tip, 10–14 mm long. FR, EP, NL, MU, YP, KI, SE; all mainland States except NT; native to Europe and the Mediterranean.

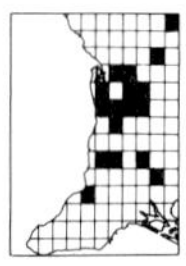

12. **Fumaria muralis*, **wall fumitory**: Scrambling annual herb with stems 30–60 cm long; lvs 5–20 cm long incl. the stalk; infl with 7–15 fls; fls (12a) pink, 9–11 mm long. FR, EP, NL, MU, YP, KI, SE; all States except Qld; native to Europe and Africa. 3 other species.

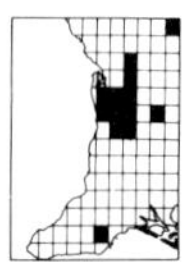

1a
1c
1b
1
2a
2
3a
3
4b
4a
4
5
6a
6b
6c
6
7a
7b
7
8b
8a
8
9b
9a
9
10a
10
11a
11
12a
12

Plate XXVII

CRUCIFERAE

1. **Brassica tournefortii*, **long-fruited wild turnip**: Annual herb to 60 cm tall; lvs usually 2–30 cm long incl. the stalk; fls (1a) pale-yellow to white, with petals 5–8 mm long; fruit (1b) 'beaked', 3–7 cm long. All regions except KI; all mainland States; native to the Mediterranean and Asia. 5 other species.

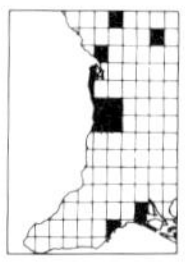

2. **Cakile maritima*, **two-horned sea-rocket**: Annual herb to 80 cm long; lvs usually 3–11 cm long; fls (2a) white to purple, 8–14 mm long; fruit (2b) 12–27 mm long; note the 2 lobes (often larger than figured) on the lower part of the fruit. Grows on coasts. NU, EP, NL, MU, YP, KI, SE; all States except NT and Qld; native to Europe. **C. edentula* lacks the lobes on the fruit.

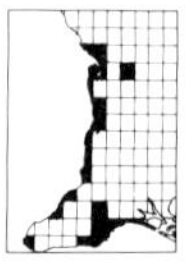

3. **Capsella bursapastoris*, **shepherds purse**: Annual herb 8–40 cm tall; basal lvs usually 3–17 cm long incl. the stalk; fls (3a) white, with petals 2–3 mm long; fruit (3b) notched, 6–9 mm long. NW, LE, FR, EA, EP, NL, MU, YP, KI, SE; all States; native to the Northern Hemisphere.

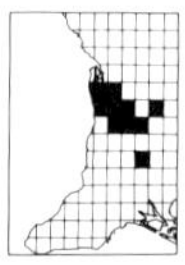

4. **Cardamine hirsuta*, **common bitter cress**: Annual herb to 30 cm tall; lvs usually to 10 cm long incl. the stalk; fls (4a) white, with petals 2.5–3.0 mm long; fruit (4b and 4c) 15–25 mm long. EP, NL, KI, SE; all States except NT; native to the Northern Hemisphere. 3 or 4 other species.

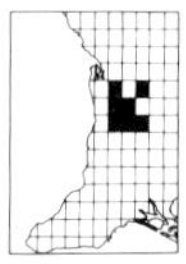

5. **Cardaria draba*, **hoary cress**: Perennial herb 15–90 cm tall; lvs usually to 10 cm long incl. the stalk; fls (5a) white, 3–4 mm long; fruit (5b) 3.0–4.5 mm long. NU, FR, NL, MU, YP, KI, SE; all States except NT; native to Asia.

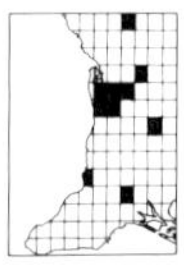

6. **Coronopus didymus*, **lesser swine cress**: Weak annual or perennial herb to 30 cm tall; lvs to 9 cm long incl. the stalk; fls (6a) white, with petals c. 0.5 mm or absent; fruit (6b) c. 1.5 mm long. FR, EA, EP, NL, MU, YP, KI, SE; all States except NT; native to S. America.

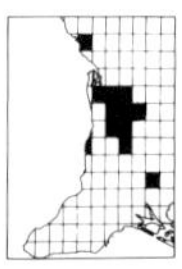

7. **Hymenolobus procumbens*, **oval purse**: Annual herb 4–30 cm tall; lvs to 2 cm long incl. the stalk; fls (7a) white, with petals to 1 mm long; fruit (7b) 3–6 mm long. NU, GT, FR, EA, EP, NL, MU, YP, KI, SE; all States except NT and Qld; New Zealand; native to Europe.

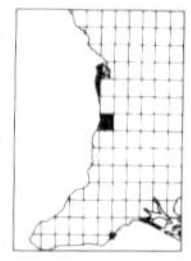

8. **Lepidium africanum*, **common peppercress**: Annual or perennial herb to 80 cm tall; lvs usually to 6 cm long; fls (8a) with petals thread-like; fruit (8b) 2–3 mm long. NU, FR, EA, EP, NL, MU, YP, SE; all States; native to S. Africa.

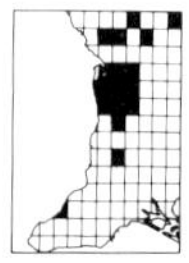

9. **Lepidium latifolium*, **dittander**: Perennial herb 50–130 cm tall; lvs to 30 cm long incl. the stalk; fls (9a) white, with petals 2–3 mm long; fruit (9b) c. 2.5 mm long. Qld, Tas.; native to Europe, Africa and Asia. 3 or 4 other species.

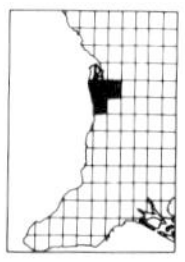

10. **Raphanus raphanistrum*, **wild radish**: Annual herb usually 15–50 cm tall; lvs to 20 cm long incl. the stalk; fls (10a) white or yellow, with petals 12–20 mm long; fruit (10b) 'beaked', 2–9 cm long. LE, FR, EP, NL, MU, YP, SE; all States; native to the ?Mediterranean.

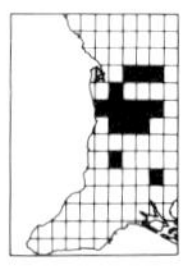

11. **Rapistrum rugosum*, **short-fruited wild turnip**: Annual or biennial herb usually 15–60 cm tall; lvs to 25 cm long incl. the stalk; fls (11a) bright yellow, 5–10 mm long; fruits (11b) with hairs (11c) without hairs 5–10 mm long. LE, FR, EA, EP, NL, YP, SE; all States except WA and NT; native to the Mediterranean.

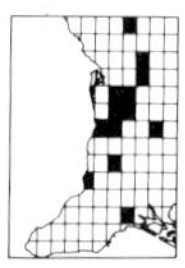

12. **Rorippa nasturtium-aquaticum*, **watercress**: Perennial herb forming mats in water, with branches to 250 cm long; lvs usually under 10 cm long incl. the stalk; fls (12a) white, with petals 3–6 mm long; fruit (12b) 10–18 mm long. FR, EP, NL, MU, SE; Qld, NSW, Vic., Tas.; native to Europe, Africa and Asia.

(*Continued*)

1
1a
1b
2
2a
2b
3
3a
3b
4
4a
4b
4c
5
5a
5b
6
6a
6b
7
7a
7b
8
8a
8b
9
9a
9b
10
10a
10b
11
11a
11b
11c
12
12a
12b

Plate XXVIII

CRUCIFERAE
(Continued)

1. **Diplotaxis muralis*, **wall rocket**: Annual or biennial herb 10-60 cm tall; lvs mostly basal, usually 5-12 cm long incl. the stalk; fls (1a) yellow, 4-9 mm long; fruit (1b) 'beaked', to 4 cm long. NU, FR, EA, EP, NL, MU, YP, KI, SE; all States except NT and Qld; native to the Mediterranean. **D. tenuifolia*, **Lincoln weed**, lacks hairs.

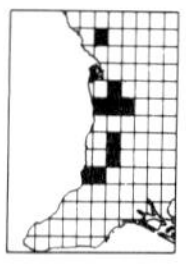

2. **Hirschfeldia incana*, **Buchan weed**: Annual or biennial herb to 1 m tall; lvs usually under 20 cm long incl. the stalk; fls (2a) yellow, 6-8 mm long; fruit (2b) 'beaked', 7-17 mm long. EP, MU, SE; all States except WA and NT; native to the Mediterranean.

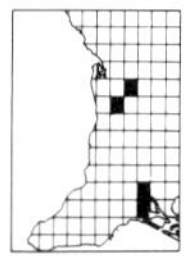

3. **Carrichtera annua*, **Wards weed**: Herb 5-40 cm tall; lvs to 2-12 cm long; fls (3a) yellow, 8-19 mm long; fruit (3b) 'beaked', 6.0-7.5 mm long. NW, LE, NU, GT, FR, EA, EP, NL, MU, YP; all States; native to the Mediterranean.

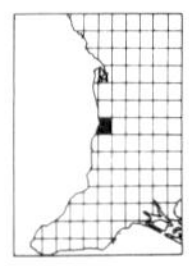

4. **Sinapis arvensis*, **charlock**: Annual herb 30-80 cm tall; lvs to 20 cm long incl. the stalk; fls (4a) yellow, 9-12 mm long; fruit (4b) 'beaked', 2.5-5.5 cm long. NL, ?KI, SE; all States; native to Europe, Africa and Asia.

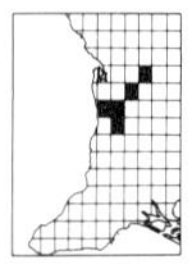

5. **Sisymbrium erysimoides*, **smooth mustard**: Annual herb 10-80 cm tall; lvs usually under 20 cm long incl. the stalk; fls (5a) yellow, 1.0-2.5 mm long; fruit (5b) straight, 2.5-5.0 cm long. All regions except KI; WA, NT, NSW, Vic.; native to the Mediterranean.

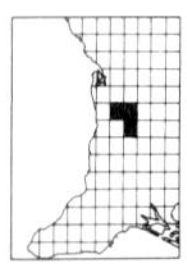

6. **Sisymbrium irio*, **London rocket:** Annual or biennial herb 25-60 cm tall; lvs usually under 18 cm long incl. the stalk; fls yellow, usually 2.5-3.5 mm long; fruit (6a) curved, 2.5-6.5 cm long; fruit stalk more slender and longer (6-10 mm) than in *S. erysimoides* (where it is 2-3 mm). LE, NU, GT, FR, EA, EP, NL, MU, YP, SE; all mainland States; native to the ?Mediterranean.

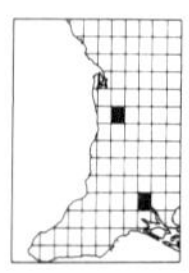

7. **Sisymbrium orientale*, **wild mustard**: Annual or biennial herb 25-90 cm tall; lvs usually under 12 cm long incl. the stalk; fls (7a) yellow, 6-10 mm long; fruit (7b) spreading, 4-10 cm long. NW, LE, GT, FR, EA, EP, NL, MU, YP, SE; all States; native to Europe, Africa and Asia. **S. officinale*, **hedge mustard**, has the fruit pressed against the stem.

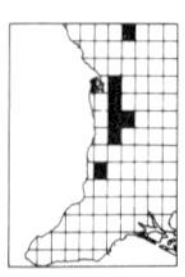

DILLENIACEAE

8. *Hibbertia exutiacies*, a **guinea-flower**: Shrub with stems on the ground, c. 30-50 cm tall; lvs (8a) sharp-tipped, 5-10 mm long; fls (8b) c. 8-10 mm across. FR, EP, NL, MU, SE; Vic.

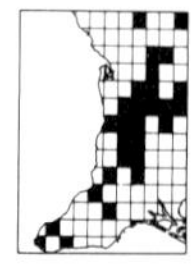

9. *Hibbertia riparia*, **erect guinea-flower**: Shrub usually under 50 cm tall; lvs (9a) 3-20 mm long; fls (9b) usually c. 1-2 cm across. FR, EP, NL, MU, YP, KI, SE; all States except NT.

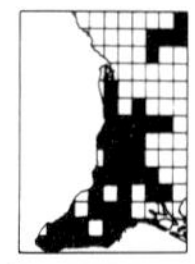

10. *Hibbertia sericea*, **silky guinea-flower**: Shrub often under 50 cm tall; lvs (10a) silky, 4-30 mm long; fls (10b) 12-30 mm across. EP, NL, MU, YP, KI, SE; all States except WA and NT.

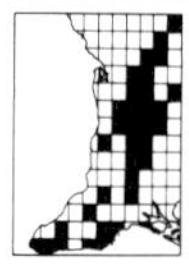

11. *Hibbertia virgata*, **twiggy guinea-flower**: Shrub often under 50 cm tall; lvs (11a) smooth or with curly hairs, 7-30 mm long; bud (11b); fls (11c) 12-25 mm across; with 10-12 stamens (more than any other species). GT, EP, NL, MU, YP, KI, SE; all States except NT and Qld. 3 other species but more are likely to be described.

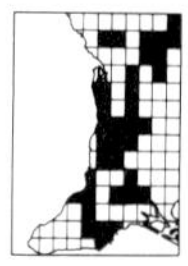

GUTTIFERAE

12. *Hypericum gramineum*, **small St Johns wort**: Annual or perennial herb 2-40 cm tall; lvs 4-26 mm long; fls (12a) with petals 3.5-10.0 mm long. LE, FR, EA, EP, NL, MU, KI, SE; all States; New Zealand, Asia. *H. japonicum*, **matted St Johns wort**, is a weaker plant with only 5-30 stamens.

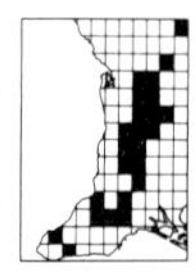

13. **Hypericum perforatum*, **St Johns wort**: Perennial herb 30-90 cm tall; lvs 1-2 cm long; fls (13a) with petals c. 10 mm long; note black dots around petals. EP, NL, MU; all States; New Zealand; native to Europe and Asia.

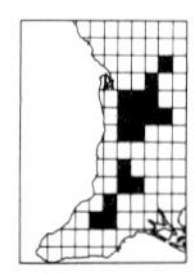

1a
1b
1
2a
2b
2
3a
3b
3
4a
4b
4
5a
5b
5
6a
6
7a
7b
7
8b
8a
8
9b
9a
9
10b
10a
10
11c
11a
11b
11
12a
12
13a
13

CRASSULACEAE

1. *Crassula colorata*, **dense crassula**: Annual herb to 15 cm tall; lvs 2-6 mm long; fls in clusters on both sides of the stem, stalkless, 1.0-2.5 mm long, with 5 petals; fruit (1a and b) 1-2 mm long. All regions except KI; all mainland States.

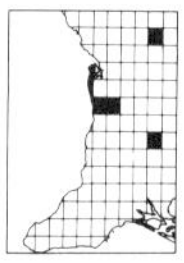

2. *Crassula decumbens*, **spreading crassula**: Annual herb with branches 0.5-12.0 cm long; lvs 3-10 mm long; fls in clusters on both sides of the stem, stalked, 1.5-3.5 mm long, with 5 petals; fruit (2a enclosed in calyx, 2b separate segment) 1.5-2.0 mm long. GT, FR, EP, NL, MU, YP, KI, SE; all States except NT and Qld; S. America, Africa.

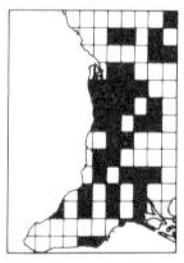

3. **Crassula alata*, **three-part crassula**: Annual herb to 5 cm tall; lvs pointed and densely clustered 1.5-3.0 mm long; fls in clusters on both sides of the stem, stalkless, 1.5-2.0 mm long, with 3 petals; fruit (3a enclosed in calyx) c. 1.5 mm long. SE; WA, Vic.; native to the Mediterranean.

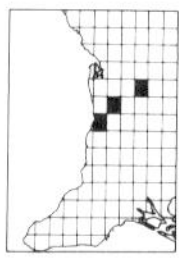

4. **Crassula natans*, a **crassula**: Annual herb with branches to 10 cm long; lvs 3-14 mm long; fls (4a) single per pair of lvs, 0.6-0.8 mm long, with 4 petals; fruit (4b in old fl., 4c separate segment) c. 1 mm long. Grows in wet places. EP, MU, YP, KI, SE; Vic., Tas.; native to S. Africa.

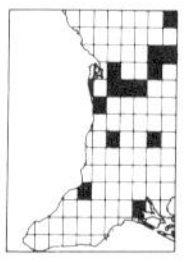

5. *Crassula helmsii*, **swamp crassula**: Annual herb with branches to 25 cm long; lvs 3-12 mm long; fls (5a) single per pair of lvs, 1.6-2.0 mm long, with 4 petals; fruit c. 1.5 mm long; note the shorter calyx in *C. natans* which has only 1 seed in each fruit segment. MU, KI, SE; NSW, Vic., Tas.; New Zealand.

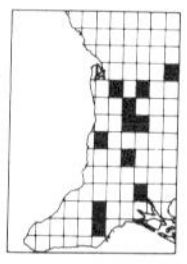

6. *Crassula sieberana*, **Sieber crassula**: Annual or perennial herb with branches to 20 cm long; lvs 3-11 mm long; fls (6a) in clusters on both sides of the stem, stalked, 1.5-3.0 mm long, with 4 petals; fruit (6b in old fl, 6c separate segment) c. 1 mm long. All regions; all States. 5 other species.

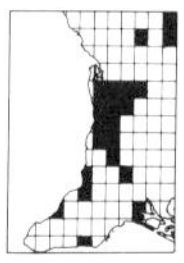

ROSACEAE

7. *Acaena agnipila*, a **sheeps burr**: Perennial herb usually 20-50 cm tall; lvs usually 8-15 cm long incl. the stalk; fls in interrupted spikes; fruit (7a) c. 3 mm long. SE; all States except NT; New Zealand.

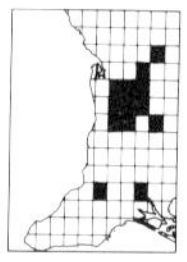

8. *Acaena echinata*, a **sheeps burr**: Perennial herb usually 25-40 cm tall; lvs usually 8-15 cm long incl. the stalk; fls in interrupted spikes; fruit (8a) 3-4 mm long; differs from *A. agnipila* in having only a few hairs on the lower surface of the lvs. FR, EP, NL, MU, YP, KI, SE; all States except NT and Qld.

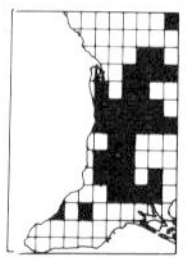

9. *Acaena novae-zelandiae*, **bidgee-widgee**: Perennial herb usually 9-25 cm tall; lvs usually 5-9 cm long incl. the stalk; fls in a single globular head; fruit (9a) 4-6 mm long excl. the spines. EP, MU, KI, SE; all States except WA and NT; New Zealand. *A. anserovina* is a hybrid between *A. novae-zelandiae* and *A. ovina*.

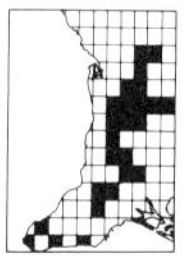

10. *Acaena ovina*, a **sheeps burr**: Perennial herb usually 10-50 cm tall; lvs usually 8-12 cm long; fls in interrupted spikes; fruit (10a) c. 4 mm long excl. the spines. YP; all States except NT.

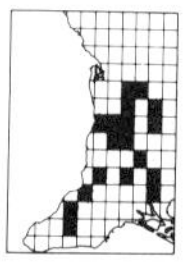

11. *Aphanes australiana*, **Australian piert**: Annual herb to 6 cm tall; lvs to c. 12 mm long; fls (11a) 1.25-2.25 mm long; fruit (11b) and persistent calyx together c. 2 mm long. EP, NL, MU, YP, KI, SE; NSW, Vic.

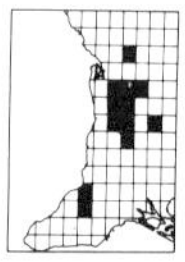

12. **Sanguisorba minor*, **sheeps burnet**: Perennial herb 10-90 cm tall; lvs 4-15 cm long incl. the stalk; fls (male 12a, female 12b) with petals c. 3 mm long; fruit (12c) 3-8 mm long. EP, NL, MU, YP, SE; all States except NT and Qld; native to Europe.

(*Continued*)

1
1a
1b
2
2a
2b
3
3a
4
4a
4b
4c
5
5a
6
6a
6b
6c
7
7a
8
8a
9
9a
10
10a
11
11a
11b
12
12a
♂
12b
♀
12c

Plate XXX

ROSACEAE
(Continued)

1. **Cotoneaster simonsii*, a **cotoneaster**: Evergreen shrub to 4 m tall; lvs 1.5–2.5 cm long; fls (1a) with petals c. 3–4 mm long; fruit (1b) c. 8 mm long. Native to India.

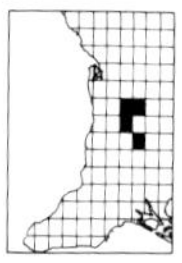

2. **Crataegus sinaica*, a **hawthorn**: Deciduous shrub or tree to 10 m tall, often spiny; lvs 2–5 cm long excl. the stalk; fls with petals 5–7 mm long; fruit 1–2 cm diam. Native to the Middle East.

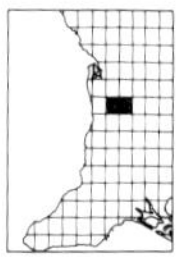

3. **Crataegus monogyna*, a **hawthorn**: Deciduous shrub or tree to 10 m tall, spiny; lvs 15–55 mm long excl. the stalk; fls 8–15 mm diam.; fruit (3a) 8–10 mm diam. NL, SE; NSW, Vic.; native to Europe.

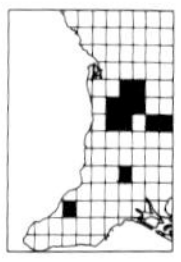

4. **Malus sylvestris*, **apple**: Deciduous tree to 10 m tall; lvs 3–11 cm long excl. the stalk; fls (4a) white or pink, 3–14 mm diam.; fruit to 10 cm diam. in wild forms. FR, MU; Vic.; native to Europe and Asia.

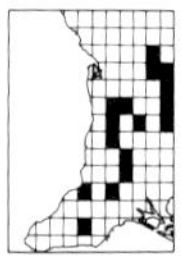

5. **Pyrus communis*, **pear**: Tree to 20 m tall; lvs (5a) 5–8 cm long excl. the stalk; fls (5b) white or pink, with petals 12–14 mm long; fruit to 10 cm long. Probably a hybrid of species native to Europe and Asia.

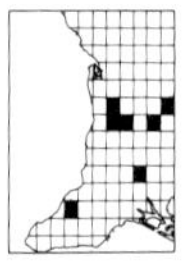

6. **Rosa canina*, **dog rose**: Prickly weak-stemmed shrub to 3 m tall; lvs with 5–7 leaflets each 15–40 mm long; fls (6a) pink, with petals 15–25 mm long; fruit (6b) 1–2 cm long. EP, NL, SE; Tas.; native to Europe.

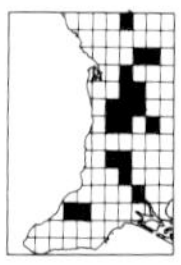

7. **Rosa rubiginosa*, **sweet briar**: Prickly weak-stemmed shrub to 3 m tall; lvs with 5–7 leaflets each 10–25 mm long; fls (7a) bright-pink, with petals 8–15 mm long; fruit (7b) 1.0–1.5 cm long; note apple-like smell when lvs crushed. EP, NL, MU, KI, SE; all States except NT; native to Europe.

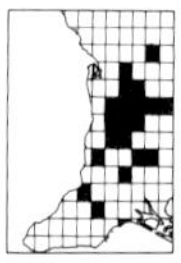

8. **Prunus armeniaca*, **apricot**: Deciduous tree to 6 m or more tall; lvs 4–7 cm long excl. the stalk; fls white or pink, with petals 10–15 mm long; fruit 3–5 cm long. NL, MU; native to ?Asia.

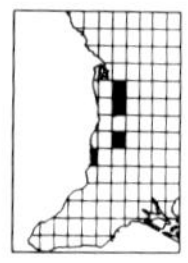

9. **Prunus cerasifera*, **cherry-plum**: Deciduous shrub or tree to 8 m tall; lvs green or purplish, 4–7 cm long excl. the stalk; fls white or pink, with petals 8–10 mm long; fruit (9a) 2–3 cm long. FR, NL, MU; Vic.; native to Europe and Asia.

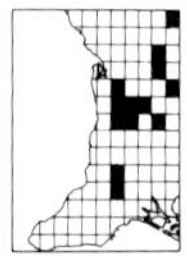

10. **Prunus domestica*, **plum**: Deciduous shrub or tree to 10 m tall; lvs 3–8 cm long excl. the stalk; fls white, with petals 7–12 mm long; fruit 2.0–7.5 cm long. FR, NL. Probably a hybrid of species native to Europe.

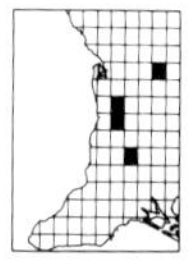

11. **Prunus dulcis*, **almond**: Deciduous shrub or tree to 8 m tall; lvs 4–12 cm long excl. the stalk; fls (11a) white, with petals c. 20 mm long; fruit 3.5–6.0 cm long. FR, EP, NL; native to Africa and Asia.

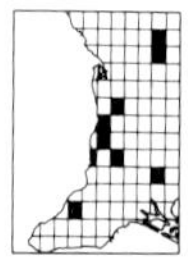

12. **Prunus persica*, **peach, nectarine**: Tree to 6 m tall; lvs 5–15 cm long excl. the stalk; fls (12a) pink or reddish-pink, with petals 10–20 mm long; fruit 4–8 cm diam. FR, NL, MU; native to Asia. **P. laurocerasus*, **cherry-laurel**, is evergreen; **P. avium*, **sweet-cherry**, has hairless fruit on stalks more than twice as long as the fruit.

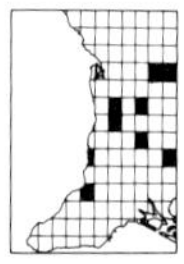

(Continued)

1a
1b
2
1
3a
3
5a
6b
4a
6a
5b
5
6
4
7
7b
7a
8
9
9a
11a
12
10
11
12a

ROSACEAE
(*Continued*)

1. 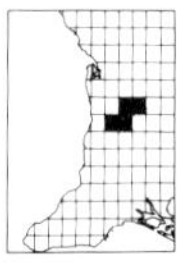 **Rubus laciniatus*, **cut-leaved blackberry**: Perennial long-stemmed shrub; lvs 5–18 cm long incl. stalk; fls (1a) white or pink, with petals 9–12 mm long; fruit (1b) c. 1 cm diam. NSW, Vic., Tas.; native to Europe.

2. 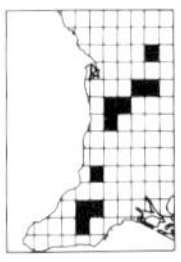 *Rubus parvifolius*, **small-leaved blackberry**: Perennial long-stemmed shrub; lvs consisting of 3 or 5 leaflets, the terminal leaflet 2–4 cm long; fls red, with petals 4–5 mm long; fruit (2a) to 10 mm diam. KI, SE; Qld, NSW, Vic.; Asia.

3. 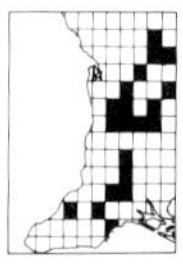 **Rubus ulmifolius*, **blackberry**: Perennial long-stemmed shrub; lvs consisting of 3 or 5 leaflets, the terminal leaflet usually 4–9 cm long; fls white, with petals 6–13 mm long; fruit c. 1 cm diam. EP, SE; Vic.; native to Europe.

PITTOSPORACEAE

4. 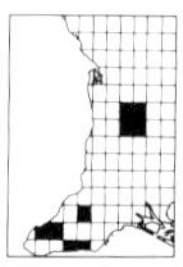 *Billardiera bignoniacea*, **orange bell-creeper**: Twining perennial to 1 m or more tall; lvs 12–85 mm long; fls (4a) greenish-yellow to red, 18–23 mm long; fruit 10–12 mm long. MU, KI; Vic.

5. 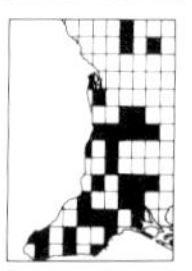 *Billardiera cymosa*, a **sweet apple-berry**: Twining perennial or shrub to 1 m or more tall; lvs 2–6 cm long; fls (5a) 11–20 mm long; fruit 8–13 mm long; like *B. versicolor* but with a stalk to the fl clusters. FR, EP, NL, MU, YP, KI, SE; Vic.

6. 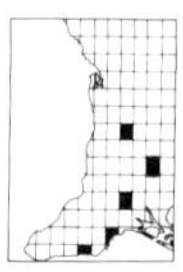 *Billardiera uniflora*, an **apple-berry**: Twining perennial to 1 m tall; lvs 14–50 mm long; fls cream or white, 12–16 mm long; fruit 3–4 cm long. EP, NL, KI.

7. 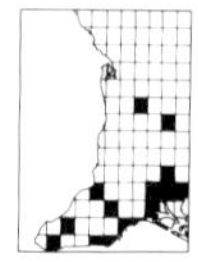 *Billardiera versicolor*, a **sweet apple-berry**: Twining perennial to 1 m or more tall; lvs 35–60 mm long; fls yellow to purple, 8–12 mm long; fruit (7a) 18–25 mm long. FR, EP, MU, YP, KI, SE; NSW, Vic. *B. sericophora*, an **apple-berry**, is similar but has grey-hairy lvs.

8. 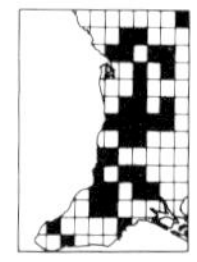 *Bursaria spinosa*, **sweet bursaria**: Shrub or tree 1–12 m tall; lvs 4–35 mm long; fls (8a) with petals 3–7 mm long; fruit (8b and 8c) 3–4 mm long. FR, EA, EP, NL, MU, YP, KI, SE; all States except WA and NT.

9. 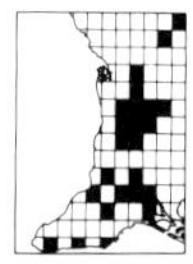 *Cheiranthera alternifolia*, **finger-flower**: Shrub to 50 cm tall; lvs 7–60 mm long; fls 12–20 mm long; fruit (9a) 12–18 mm long. FR, EP, NL, MU, YP, KI, SE; Vic.

10. 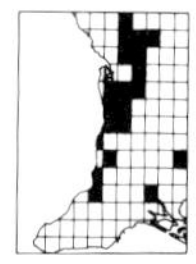 *Pittosporum phylliraeoides*, **native apricot**: Tree to 17 m tall, with drooping branches; lvs 25–125 mm long; fls (10a) yellow to white, with petals 7–12 mm long; fruit (10b) 10–24 mm long. All regions; all mainland States.

11. 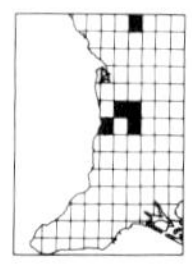 **Pittosporum undulatum*, **sweet pittosporum**: Shrub or tree 3–13 m tall; lvs 6–17 mm long; fls (11a) with petals 11–17 mm long; fruit (11b) 8–14 mm long. Native to Qld, NSW, Vic., Tas.

12. 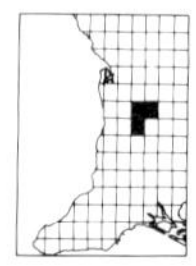 **Sollya heterophylla*, **sollya**: Twining perennial or shrub to 2 m tall; lvs 16–60 mm long; fls (12a) with petals 4–12 mm long; fruit (12b) 1.5–2 cm long. Tas.; native to WA.

1b
1
1a
2
2a
3
4a
4
5
5a
6
7
7a
8
8a
8b
8c
9
9a
10a
10b
10
11a
11b
11
12b
12a
12

Plate XXXII

LEGUMINOSAE

1. *Acacia dodonaeifolia*, **sticky wattle**: Somewhat sticky shrub or tree 2–6 m tall; 'lvs' 4–10 cm long; fruit (1a) 6–10 cm long. EP, YP, KI.

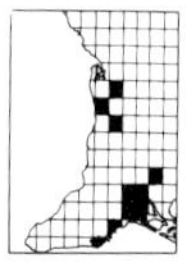

2. *Acacia gunnii*, **ploughshare wattle**: Prickly shrub 20–80 cm tall; 'lvs' (2a) 5–13 mm long; fruit (2b) 2–3 cm long; high rainfall areas. All States except WA and NT.

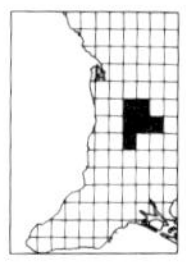

3. *Acacia ligulata*, **umbrella bush**: Shrub 1–5 m tall (lower along coast); 'lvs' 3–9 cm long; fruit (3a) 5–10 cm long, easily breaking up at the constrictions. All regions; all States except Tas.

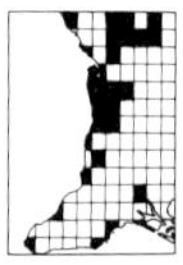

4. *Acacia melanoxylon*, **blackwood**: Tree to 15 m or more tall; 'lvs' 6–14 cm long, the true compound lvs present on young plants; fruit (4a) 4–12 cm long; timber used for furniture. EP, NL, SE; all States except WA and NT.

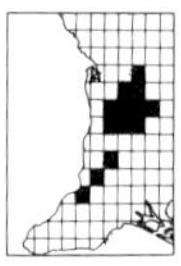

5. *Acacia myrtifolia*, **myrtle wattle**: Shrub 1–2 m tall; 'lvs' 2–10 cm long; fruit (5a) 4–7 cm long. NL, YP, KI, SE; all States except NT.

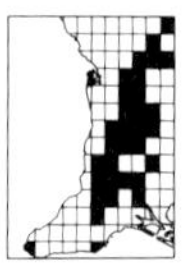

6. *Acacia notabilis*, **notable wattle**: Shrub 1–3 m tall; 'lvs' thick, 5–15 cm long; fruit (6a) 3–7 cm long. FR, EP, NL, MU, YP, SE; NSW, Vic.

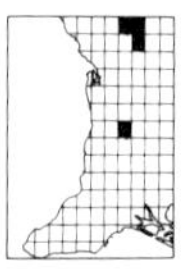

7. *Acacia paradoxa*, **kangaroo thorn**: Prickly shrub to 3 m tall; 'lvs' (7a) 1–2 cm long; fruit (7b) 4–7 cm long. FR, EA, EP, NL, MU, YP, KI, SE; all States except NT.

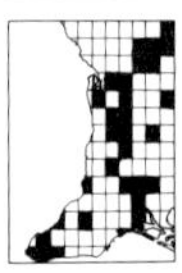

8. *Acacia pycnantha*, **golden wattle**: Shrub or tree c. 4–8 m tall; bark dark brown to pale grey; 'lvs' 6–20 cm long; fls fragrant; fruit (8a) 5–12 cm long. FR, EA, EP, NL, MU, YP, KI, SE; NSW, Vic.

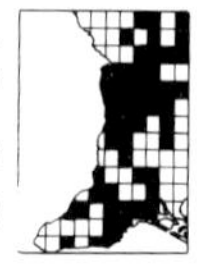

9. *Acacia retinodes*, **wirilda**: Shrub or tree to 8 m tall; 'lvs' 3–20 cm long; fruit (9a) 3–14 cm long; mainly in damp areas. EP, NL, MU, KI, SE; Vic., Tas.

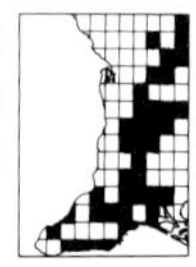

10. *Acacia acinacea*, **round-leaved wattle**: Shrub to 2 m tall; 'lvs' (10a) 5–14 mm long; fruit (10b) c. 2–5 cm long but often tightly coiled. EP, NL, YP, KI, SE; NSW, Vic.

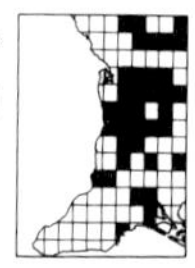

11. *Acacia verniciflua*, **varnish wattle**: Sticky shrub or small tree 3–5 m tall; 'lvs' 4–8 mm long; fruit (11a) 6–10 cm long; high rainfall areas. All States except WA and NT.

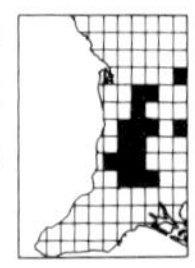

12. **Acacia baileyana*, **Cootamundra wattle**: Shrub or tree to 6 m tall; lvs 2–5 cm long excl. the stalk; fruit (12a) 4–10 cm long. NL; Qld, Vic.; native to NSW.

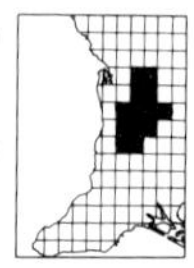

(*Continued*)

1a
1
2b
2
2a
3
3a
4
4a
5
5a
6
6a
7
7a
7b
8
8a
9
9a
10
10a
10b
11
11a
12
12a

Plate XXXIII

LEGUMINOSAE

(*Continued*)

1. *Acacia calamifolia*, **wallowa**: Shrub 2-4 m tall; 'lvs' 4-9 cm long; fruit (1a) 7-15 cm long. FR, EA, EP, NL, MU, YP, SE; NSW, Vic.

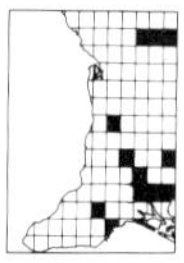

2. *Acacia continua*, **thorn wattle**: Prickly shrub to 1 m tall; 'lvs' 2-4 cm long; fruit (2a) 3-8 cm long. FR, EA, EP, NL, MU; NSW.

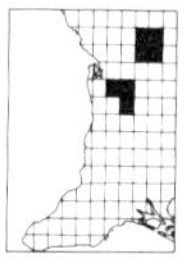

3. *Acacia rupicola*, **rock wattle**: Prickly shrub 1.0-2.5 m tall; 'lvs' 1.5-2.5 cm long; fruit (3a) 4-7 cm long. FR, EP, NL, MU, YP, KI, SE; Vic.

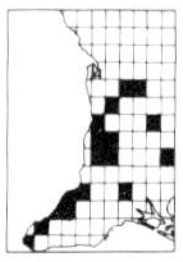

4. *Acacia longifolia*, **sallow wattle**: Shrub 3-7 m tall; 'lvs' 6-15 cm long; fruit (4a) 5-10 cm long. EP, YP, KI, SE; all States except WA and NT.

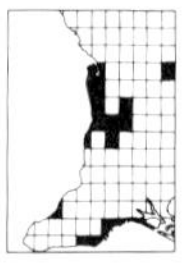

5. *Acacia verticillata*, **prickly Moses**: Shrub 1-4 m tall; 'lvs' 8-15 mm long; fruit (5a) 3-8 cm long; usually in damp places. FR, KI, SE; NSW, Vic., Tas. 20 other species.

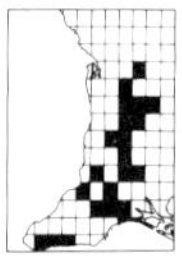

6. *Senna artemisioides*, **punty bush**: Shrub or small tree 1-3 m tall; each of the 2 leaflets 2-5 cm long; fls (6a) c. 15 mm across; fruit 2-8 cm long. All regions except KI; all mainland States.

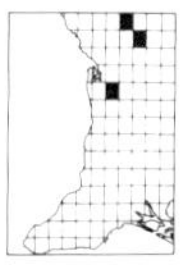

7. **Ulex europaeus*, **gorse**: Shrub 60-200 cm tall; 'lvs' 12-30 mm long; fls (7a) 15-20 mm long; fruit 1-2 cm long. MU, NL, SE; all States except NT; native to Europe.

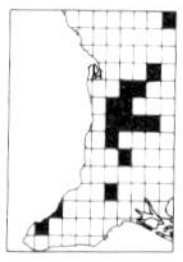

8. **Cytisus proliferus*, **tree lucerne**: Shrub or tree to 3 m or more tall; each of the 3 leaflets 1-3 cm long; fls (8a) c. 15 mm long; fruit 4-5 cm long. SE; all States except NT and Qld; native to the Canary Islands.

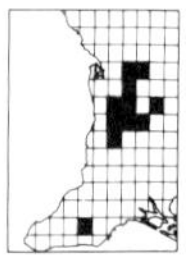

9. **Cytisus scoparius*, **English broom**: Shrub to 2 m or more tall; each of the 3 leaflets 6-20 mm long; fls (9a) 16-20 mm long; fruit 2.5-4.0 cm long. NL; NSW, Vic., Tas.; New Zealand; native to N. Africa and Europe.

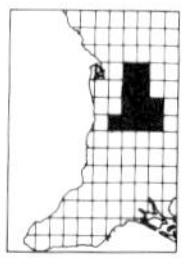

10. **Spartium junceum*, **Spanish broom**: Shrub to 3 m or more tall; lvs 1-3 cm long; fls (10a) 2-3 cm long; fruit 6-8 cm long; note the simple lvs. EP; Native to the Mediterranean.

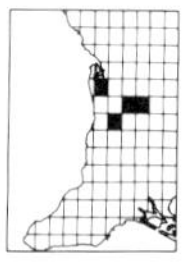

11. **Genista monspessulana*, **Montpellier broom**: Shrub 1-3 m tall; each of 3 leaflets 5-13 mm long; fls (11a) 10-13 mm long; fruit 15-25 mm long. NL, MU, SE; NSW, Vic., Tas.; New Zealand; native to the Mediterranean. **G. linifolia*, **flax-leaved broom**, has narrower lvs and the largest petal hairy.

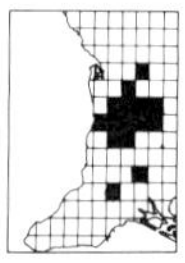

12. *Viminaria juncea*, **golden spray**: Shrub to 5 m tall; lvs very slender, 1-5 cm or more long; fls (12a) 7-10 mm long; fruit c. 4 mm long; occurs in damp places. KI, SE; all States except NT.

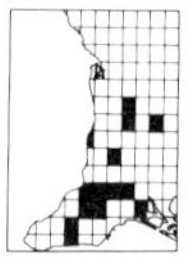

(*Continued*)

1
1a
2
2a
3
3a
4
4a
5
5a
6
6a
7
7a
8
8a
9
9a
10
10a
11
11a
12
12a

Plate XXXIV

LEGUMINOSAE
(*Continued*)

1. ******Dipogon lignosus*, **lavatory creeper**: Twining perennial herb with stems to 3 m long; each of the 3 leaflets 1–7 cm long; fls (1a) c. 1 cm long; fruit 2.5–5.0 cm long. EP, YP, KI, SE; WA, Qld, NSW, Vic.; native to S. Africa.

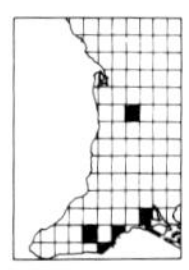

2. *Hardenbergia violacea*, **native lilac**: Twining shrub with stems to 2 m or more long; lvs 3–10 cm long excl. the stalk; fls c. 1 cm long; fruit c. 8 mm long. FR, EA, EP, NL, MU, YP, KI, SE; all States except WA and NT.

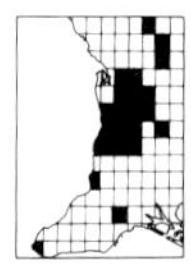

3. *Indigofera australis*, **austral indigo**: Shrub to 2 m tall; lvs 1.5–10 cm long; fls (3a) 7–10 mm long; fruit 2–4 cm long. FR, EA, EP, NL, SE; all States.

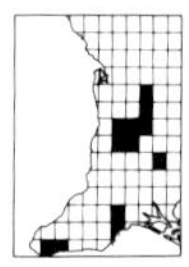

4. **Lathyrus latifolius*, **perennial pea**: Twining perennial herb with stems 0.5–3.0 m long; each of the 2 leaflets 3–15 cm long; infl of 3–15 fls; fls (4a) 2–3 cm long; fruit 5–10 cm long. ?NL; Vic., Tas.; native to Europe.

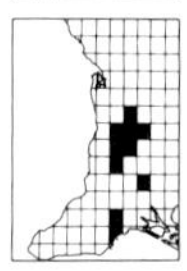

5. **Lathyrus tingitanus*, **Tangier pea**: Twining annual herb with stems 0.6–1.2 m long; each of the 2 leaflets 2–8 cm long; infl of 1–3 fls; fls (5a) 2–3 cm long; fruit 6–10 cm long. WA; native to Europe. **L. sphaericus*, **slender wild-pea**, does not have winged stems.

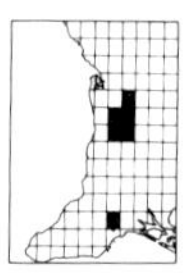

6. **Lupinus cosentinii*, **Western Australian blue lupin**: Annual herb 20–120 cm tall; each of the 7–11 leaflets 2.5–6.0 cm long; fls (6a) 12–17 mm long; fruit 4.0–5.5 cm long. EP, NL, SE; WA, NSW; native to the Mediterranean.

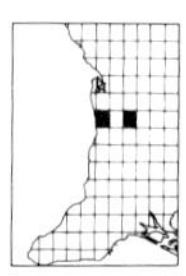

7. *Psoralea australasica*, **native scurf-pea**: Shrub 0.5–2.5 m tall; lvs 2–10 cm long excl. the stalk; fls (7a) pink or bluish, 4–8 mm long; fruit 3.0–3.5 mm long. NW, LE, GT, FR, EA, EP, MU, YP; all mainland States. 2 other species.

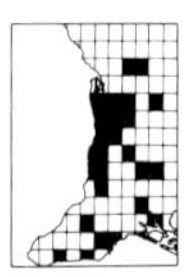

8. *Swainsona lessertiifolia*, **coast swainson-pea**: Herb 10–30 cm tall; lvs usually 6–9 cm long incl. the stalk; fls purplish or white, 8–10 mm long; fruit 1.5–3.0 cm long. GT, EP, MU, YP, KI, SE; NSW, Vic., Tas.

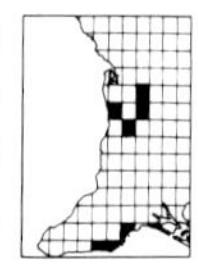

9. *Swainsona oroboides*, **variable swainson-pea**: Herb 5–30 cm tall; lvs 2–12 cm long incl. the stalk; fls bluish, purple or red, 7–9 mm long; fruit 8–20 mm long. NW, LE, NU, GT, EP, NL, MU, SE; all mainland States.

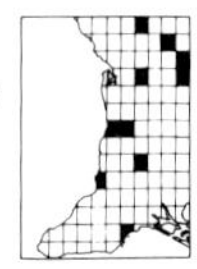

10. **Vicia hirsuta*, **hairy vetch**: Weak annual herb with stems 20–80 cm long; lvs 3–8 cm long incl. stalks; fls (10a) whitish or pale-blue, 3–4 mm long; fruit (10b) 6–11 mm long, sometimes almost hairless. SE; all mainland States; native to Europe, Asia and Africa.

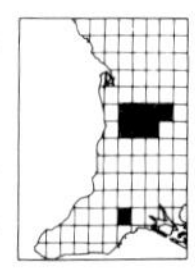

11. **Vicia monantha*, **spurred vetch**: Weak annual herb with stems 5–80 cm long; lvs 2–7 cm long incl. stalks; fls stalked violet to blue, 12–20 mm long; fruit (11a) 3–50 mm long. NW, LE, FR, EA, EP, NL, MU, YP; all States; native to Europe, Asia and Africa.

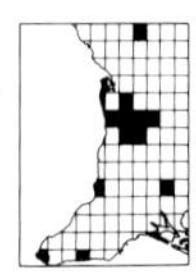

12. **Vicia sativa*, **common vetch**: Weak annual herb with stems 20–100 cm long; lvs 2–10 cm long incl. stalks; fls not stalked, purple or red to white, 9–30 mm long; fruit (12a) 2.5–8.0 cm long. EA, EP, NL, YP, KI, SE; all mainland States; native to Europe and ?Asia.

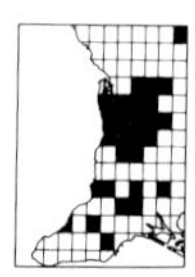

(*Continued*)

1
1a
2
3
3a
4
4a
5
5a
6
6a
7
7a
8
9
10
10a
10b
11
11a
12
12a

PLATE XXXV

LEGUMINOSAE
(*Continued*)

1. *Bossiaea prostrata*, **creeping bossiaea**: Weak creeping shrub with stems 20–30 cm long; lvs 1–2 cm long excl. the stalk; fls 8–11 mm long; fruit 2–3 cm long. EP, SE; all States except WA and NT. *B. walkeri*, **cactus pea**, has flattened stems and no lvs.

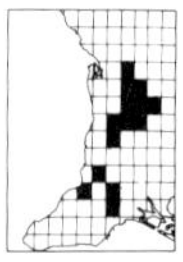

2. *Daviesia brevifolia*, **leafless bitter-pea**: Shrub to 1 m tall; lvs usually only in the form of 2–5 long spines; fls 5–6 mm long; fruit (2a) c. 15 mm long. NU, EP, NL, MU, KI, SE; NSW, Vic.

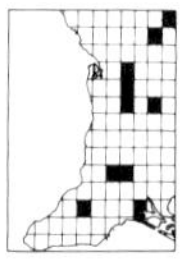

3. *Daviesia leptophylla*, a **bitter-pea**: Shrub 1–2 m tall; 'lvs' 3–10 cm long; fls 5–7 mm long; fruit (3a) c. 2.5 mm long. FR, EP, NL, MU, KI; Qld, NSW, Vic.

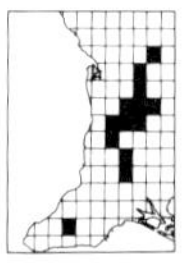

4. *Daviesia ulicifolia*, **gorse bitter-pea**: Shrub 1–2 m tall; 'lvs' 10–25 mm long; fls 4–8 mm long; fruit 7–8 mm long. NW, NU, GT, FR, NL, EP, MU, YP, KI, SE; all States except NT. 5 other species.

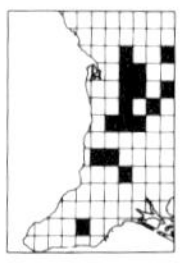

5. *Dillwynia hispida*, **red parrot-pea**: Shrub to 50 cm tall; lvs usually 3–10 mm long; fls 7–12 mm long; fruit c. 4 mm long. EP, NL, MU, KI, SE; NSW, Vic. *D. glaberrima*, **smooth parrot-pea**, also has fls in stalked clusters, but is virtually hairless and has a yellow to reddish lower lip to the fl.

6. *Dillwynia sericea*, **showy parrot-pea**: Shrub 0.3–1.0 m tall; lvs 5–20 mm long; fls 6–10 mm long; fruit c. 5 mm long. KI, SE; all States except WA and NT.

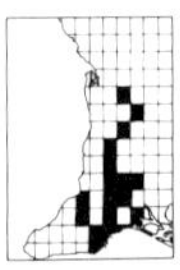

7. *Glycine clandestina*, **twining glycine**: Stem twining, 10–100 cm long; each of the 3 leaflets 6–40 mm long; fls 6–7 mm long; fruit 12–30 mm long. NW, LE, NU, GT, FR, EP, NL, MU, YP, SE; all mainland States. *G. latrobeana*, **clover glycine**, stems do not twine.

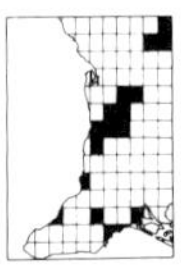

8. *Gompholobium ecostatum*, **dwarf wedge-pea**: Shrub to 50 cm tall; each of the 3 leaflets to 10 mm long; fls 15–20 mm long, yellowish to reddish; fruit 10–12 mm long. EP, KI, SE; NSW, Vic., Tas.

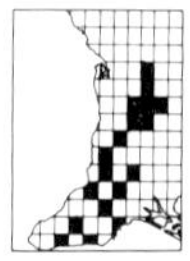

9. *Kennedia prostrata*, **running postman**: Perennial with weak sometimes twining stems to 1.5 m long; each of the 3 leaflets 1–3 cm long; fls 2.0–2.5 cm long; fruit 3–5 cm long. FR, EP, NL, YP, KI, SE; all States except NT. **K. nigricans*, **black coral-pea**, has black fls.

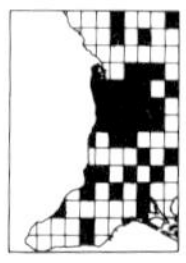

10. *Lotus australis*, **Australian trefoil**: Perennial herb to 60 cm tall; each of 5 leaflets 1–4 cm long; fls pink or white, 1–2 cm long; fruit 3–5 cm long. FR, EA, EP, NL, MU, YP, KI, SE; all States. *L. cruentus*, **redflower lotus**, has smaller lvs and usually reddish fls.

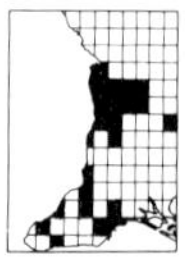

11. **Lotus uliginosus*, **birds-foot trefoil**: Perennial herb with stems 20–80 cm long; each of 5 leaflets 8–25 mm long; fls 10–18 mm long; fruit 2–3 cm long. EP, NL, MU, SE; all States except NT and Qld; New Zealand; native to Europe. **L. subbiflorus*, **hairy birds-foot trefoil**, has smaller yellow fls.

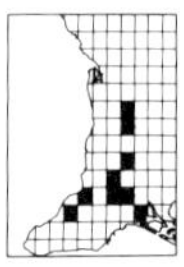

12. **Medicago sativa*, **lucerne**: Perennial herb 30–120 cm tall; each of 3 leaflets 5–25 mm long; fls 5–12 mm long; fruit (12a) 3–9 mm diam. FR, MU; all States at least in cultivation; native to ?Europe and ?Asia. Other species on following plate.

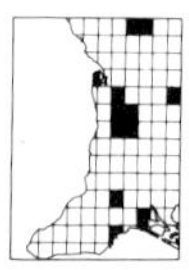

(*Continued*)

1
2
2a
3
3a
4
5
6
7
8
9
10
11
12
12a

Plate XXXVI

LEGUMINOSAE
(Continued)

1. **Medicago arabica*, **spotted medic**: Annual herb with branches 20–60 cm long; each of 3 leaflets 10–25 mm long; fls 4–5 mm long; fruit (1a) 4–6 mm broad. MU, YP, SE; Tas.; New Zealand; native to the Northern Hemisphere.

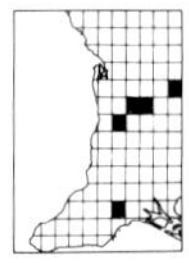

2. **Medicago lupulina*, **black medic**: Perennial herb with branches 5–60 cm long; each of 3 leaflets 11–14 mm long; fls 2.5–3.5 mm long; fruit (2a) c. 2.5 mm long. SE; all States except NT and ?Qld; New Zealand; native to Europe and Asia.

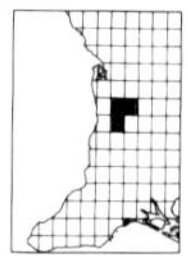

3. **Medicago minima*, **woolly burr-medic**: Annual herb with branches 5–90 cm long; each of 3 leaflets 6–14 mm long; fls 3–6 mm long; fruit (3a) 3–5 mm diam. FR, EA, EP, NL, MU, YP, SE; all States except ?Qld; native to Europe.

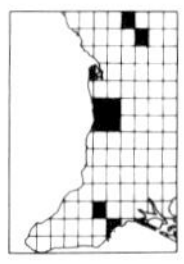

4. **Medicago polymorpha*, **burr-medic**: Annual herb with branches to 40 cm long; each of 3 leaflets 10–20 mm long; fls 4–6 mm long; fruit (4a) 4–10 mm diam. LE, NU, GT, FR, EA, EP, YP, SE; all States; native to the Mediterranean.

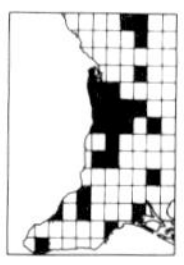

5. **Medicago scutellata*, **snail medic**: Annual herb with branches 20–60 cm long; each of 3 leaflets 15–30 mm long; fls 7–9 mm long; fruit (5a) 9–18 mm diam. YP; all States except NT and Qld; native to the Mediterranean.

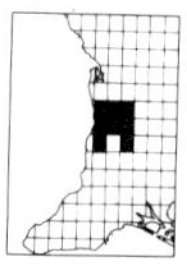

6. **Medicago truncatula*, **barrel medic**: Annual herb with branches to 50 cm long; each of 3 leaflets 7–20 mm long; fls 5–8 mm long; fruit (6a) 4.5–8 mm diam. NU, FR, EP, MU, YP, KI, SE; WA, NT, NSW, Vic.; native to the Mediterranean. 4 or 5 other species; see also previous plate.

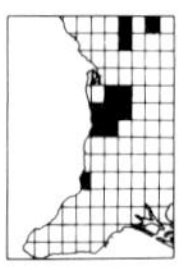

7. *Eutaxia microphylla*, **eutaxia**: Shrub often c. 50 cm tall; lvs opposite, usually hairless, 2–10 mm long; fls 6–10 mm long; note absence of paired brown stipules at the base of the lvs thus distinguishing it from *Pultenaea*. NU, FR, EP, NL, MU, YP, KI, SE; all States except NT.

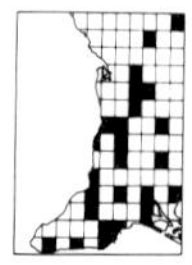

8. *Goodia lotifolia*, **golden-tip**: Shrub 1–3 m tall; each of 3 leaflets 10–25 mm long; fls 3–6 mm long; fruit 2–3 cm long. FR, EA, EP, MU, YP, KI, SE; all States except NT and Qld.

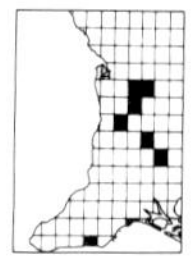

9. **Melilotus indica*, **King Island melilot**: Annual herb 5–50 cm tall; each of 3 leaflets 12–25 mm long; fls (9a) to 3 mm long; fruit 2–3 mm long. FR, EP, NL, MU, YP, KI, SE; all States except ?Qld; native to the Mediterranean and Asia.

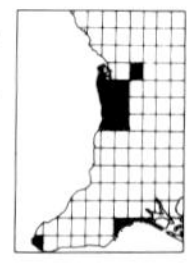

10. *Phyllota pleurandroides*, **heathy phyllota**: Shrubs to 50 cm tall; lvs 6–10 mm long; fls (10a) 5–8 mm long; note the alternate lvs separate it from *Eutaxia*. MU, KI, SE; Vic.

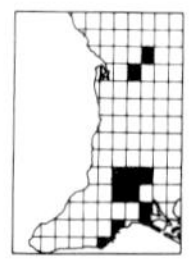

11. *Platylobium obtusangulum*, **common flat-pea**: Shrub with wiry branches to 1 m long; lvs prickly, 2–3 cm long; fls 12–15 mm long; fruit (11a) 1.5–3.0 cm long. MU, KI, SE; Vic., Tas.

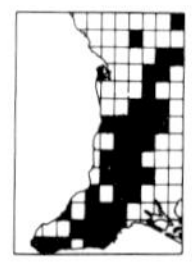

12. *Sphaerolobium vimineum*, **leafless globe-pea:** Shrub 20–70 cm tall; lvs usually less than 5 mm long; fls 4–6 mm long; fruit (12a) c. 3 mm diam. EP, SE; all States except NT.

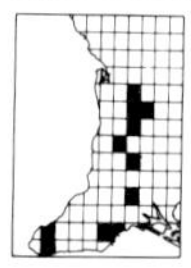

(Continued)

1a
1
2a
2
3a
3
4a
4
5a
5
6a
6
7
8
9a
9
10a
10
11
11a
12a
12

Plate XXXVII

LEGUMINOSAE
(Continued)

1. *Pultenaea acerosa*, **bristly bush-pea**: Shrub 30–60 cm tall; lvs 5–10 mm long; fls (1a and 1b) 6–9 mm long. EP, YP, KI, SE; Vic.

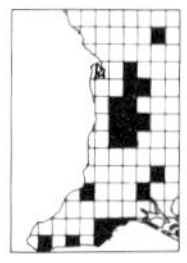

2. *Pultenaea canaliculata*, **coast bush-pea**: Shrub 1–2 m tall; lvs 8–12 mm long; fls 7–8 mm long. EP, YP, KI, SE; Vic.

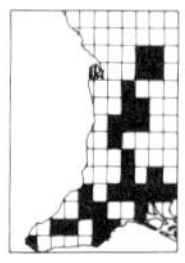

3. *Pultenaea daphnoides*, **large-leaved bush-pea**: Shrub 1–3 m tall; lvs 1–3 cm long; fls 7–15 mm long. EP, MU, KI; all States except WA and NT.

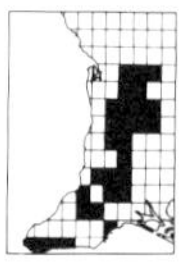

4. *Pultenaea hispidula*, **rusty bush-pea**: Shrub to 1 m tall; lvs 3–10 mm long; fls 5–8 mm long. FR, SE; Qld, NSW, Vic.

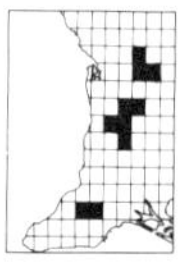

5. *Pultenaea involucrata*, **Mount Lofty bush-pea**: Shrub to 1 m tall; lvs 8–12 mm long; fls c. 8 mm long; note the bracts (5a) at the base of the fls. Restricted to Mount Lofty region.

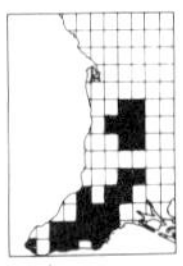

6. *Pultenaea largiflorens*, **twiggy bush-pea**: Shrub often 1 m or more tall; lvs 4–10 mm long; fls 6–8 mm long. FR, EP, SE; NSW, Vic.

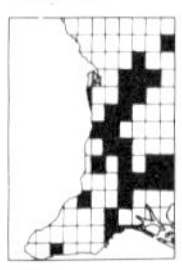

7. *Pultenaea laxiflora*, **loose-flower bush-pea**: Shrub with branches to 50 cm along the ground; shortly hairy when young; lvs 4–12 mm long; fls c. 7 mm long. KI, SE; NSW, Vic.

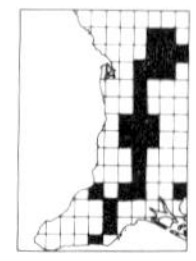

8. *Pultenaea pedunculata*, **matted bush-pea**: Shrub with branches along the ground; lvs 5–12 mm long; fls 4–8 mm long. EP, SE; NSW, Vic., Tas.

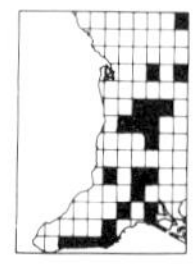

9. *Pultenaea scabra*, **rough bush-pea**: Shrub 1–2 m tall; lvs 5–15 mm long; fls 7–12 mm long. KI, SE; NSW, Vic.

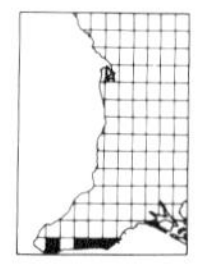

10. *Pultenaea tenuifolia*, a **bush-pea**: Shrub to 1 m tall; lvs usually 4–8 mm long; fls (10a) usually 5–8 mm long. EP, MU, YP, KI, SE; WA, Vic., Tas.

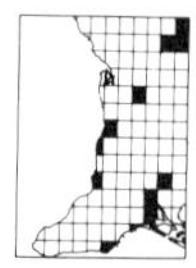

11. *Pultenaea trinervis*, **three-nerved bush-pea**: Shrub to 1 m tall; lvs (11a) 8–12 mm long; fls c. 8 mm long. EP, KI.

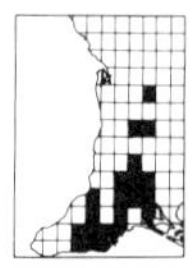

12. *Pultenaea viscidula*, **dark bush-pea**: Shrub to 1 m tall; young branches sticky; lvs 8–14 mm long; fls 6–8 mm long. KI. 5 other species.

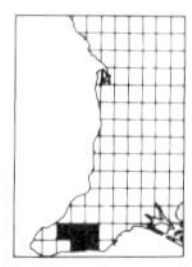

(Continued)

1
1a
1b
2
3
4
5
5a
6
7
8
9
10
10a
11
11a
12

Plate XXXVIII

LEGUMINOSAE

(Continued)

1. **Trifolium arvense*, **hares foot clover**: Annual or biennial herb with stems 4–40 cm long; infl to 3 cm long; fls white and pink; calyx (1a) 5–6 mm long. FR, EP, MU, YP, SE; all States except NT; New Zealand; native to Europe.

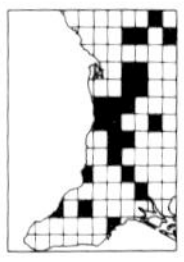

2. **Trifolium angustifolium*, **narrow-leaved clover**: Annual herb with stems 8–60 cm long; infl 15–80 mm long; fls purple to white; calyx (2a) to 10 mm long. FR, EP, NL, MU, YP, KI, SE; all States except NT and Qld; native to Europe.

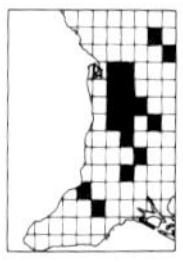

3. **Trifolium campestre*, **hop clover**: Annual herb 4–50 cm tall, hairy or not; infl 8–15 mm broad; fls yellow; calyx 1.5–2.0 mm long. FR, EP, NL, MU, YP, KI, SE; all States; New Zealand; native to Europe.

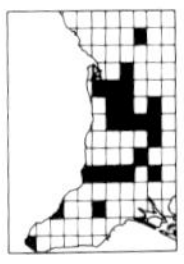

4. **Trifolium cherleri*, **cupped clover**: Annual herb with stems 5–30 cm long; infl 8–16 mm broad; fls reddish to cream; calyx 6–8 mm long, usually longer than the petals. WA; native to the Mediterranean.

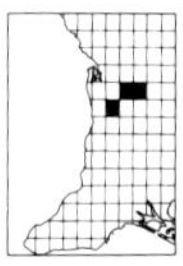

5. **Trifolium dubium*, **suckling clover**: Annual herb with stems 20–40 cm long; infl 8–9 mm long; fls yellow; calyx 1.5–2.0 mm long. EP, KI, SE; all States except NT and ?Qld; New Zealand; native to Europe.

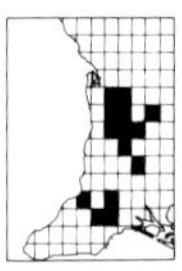

6. **Trifolium repens*, **white clover**: Perennial herb with stems 10–30 cm long; infl 15–25 mm broad; fls white or pink; calyx 2–6 mm long; used in lawns. MU, SE; all States except NT; native to Europe.

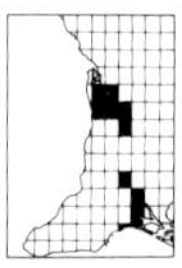

7. **Trifolium glomeratum*, **cluster clover**: Annual herb with stems 2–35 cm long; infl 8–12 mm broad; fls pink; calyx 2–3 mm long. LE, FR, EP, NL, MU, YP, KI, SE; all States except NT and Qld; New Zealand; native to Europe.

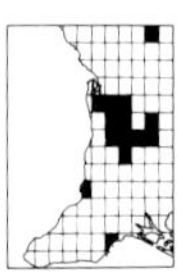

8. **Trifolium pratense*, **red clover**: Perennial herb with stems usually 20–40 cm long; infl 2–4 cm broad; fls reddish-purple to white; calyx c. 6 mm long. SE; all States except NT and ?Qld; native to Europe.

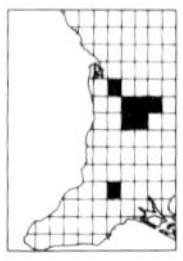

9. **Trifolium resupinatum*, **shaftal clover**: Annual herb with stems 10–60 cm long; infl 10–15 mm broad in flower; fls pink or reddish-purple; calyx (9a) c. 2 mm long but enlarging and inflating in fruit. EP, SE; all States except NT and Qld; New Zealand; native to Europe.

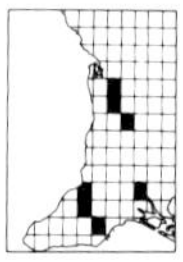

10. **Trifolium scabrum*, **rough clover**: Annual herb with stems 5–25 cm long; infl 5–20 mm broad; fls white or pink; calyx to 8 mm long. YP, SE; all States except NT and Qld; New Zealand; native to Europe.

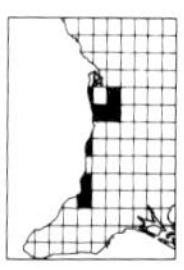

11. **Trifolium subterraneum*, **subterranean clover**: Annual herb with stems 2–30 cm long; fls whitish, 8–14 mm long; calyx 3–4 mm long; infl forced into the ground during fruiting. EP, YP, KI, SE; all States except NT and ?Qld; native to Europe and the Mediterranean. 7 other species.

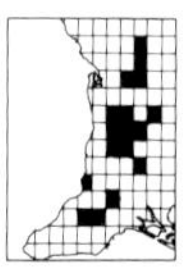

12. **Trifolium tomentosum*, **woolly clover**: Annual herb with stems 10–60 cm long; infl 6–8 mm broad; fls pink; calyx woolly, c. 3.5 mm long, but enlarging and inflating in fruit. GT, FR, EP, NL, MU, KI, SE; all States except NT; New Zealand; native to the Mediterranean. 7 other species.

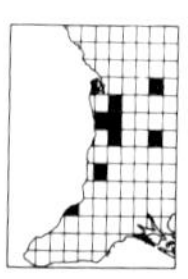

(Continued)

1
1a
2
2a
3
4
5
6
7
8
9
9a
10
11
12

OXALIDACEAE

1. *Oxalis perennans*, **native oxalis**: Herb erect or creeping, rarely to 30 cm tall; leaflets 3, each 2–15 mm long; fls (1a) with yellow petals 6–12 mm long; fruit (1b) 4–30 mm long. All regions except GT; all mainland States; New Zealand.

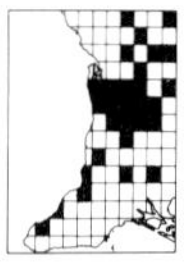

2. **Oxalis pes-caprae*, **soursob**: Perennial herb to 40 cm tall, with underground bulbils; leaflets 3, each 1–3 cm long; fls (2a) with yellow petals 15–25 mm long; fruit c. 8 mm long. FR, EP, MU, YP, KI, SE; all mainland States; native to S. Africa.

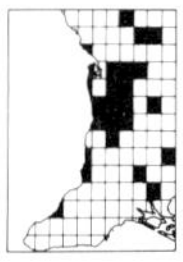

3. **Oxalis purpurea*, **one o'clock**: Perennial herb to 30 cm tall, with underground bulbils; leaflets 3, each 1–3 cm long; fls with pink to purple rarely white petals 15–40 mm long. EP, NL, MU, YP, KI; all States except NT and Qld; native to S. Africa. 11 other species.

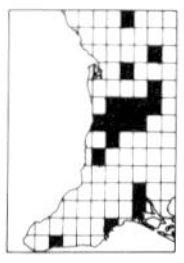

GERANIACEAE

4. **Erodium botrys*, **long storks bill**: Annual herb to 40 cm tall; lvs to 10 cm long excl. the stalk; fls (4a) with petals to 2 cm long; fruit (4b awn incomplete) c. 10 mm long excl. awn. FR, EP, NL, MU, YP, KI, SE; all mainland States; native to the Mediterranean.

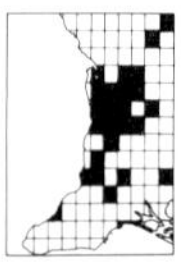

5. **Erodium cicutarium*, **common storks bill**: Annual or biennial herb with stems to 40 cm long; lvs to 7 cm long excl. the stalk; fls (5a) with petals to 1 cm long; fruit (5b awn incomplete) c. 8 mm long excl. awn. LE, NU, GT, FR, EA, EP, NL, MU, YP, KI, SE; all States; native to Europe.

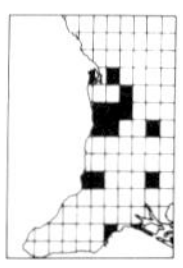

6. *Erodium crinitum*, **blue storks bill**: Herb with weak stems to 50 cm long; lvs to 4 cm long excl. the stalk; fls (6a) with petals to 15 mm long; fruit (6b awn incomplete) c. 7 mm long excl. awn. NW, LE, NU, GT, FR, EA, EP, NL, MU, YP, SE; all States.

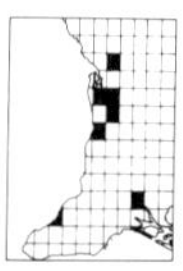

7. **Erodium moschatum*, **musky storks bill**: Annual herb with weak stems to 50 cm long; lvs to 15 cm long excl. the stalk; fls (7a) with petals to 10 mm long; fruit (7b) to 9 mm long excl. awn. EP, NL, MU, YP, KI, SE; NSW, Vic.; native to the Mediterranean. 2 other species.

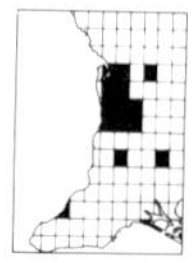

8. *Geranium potentilloides*, a **geranium**: Perennial herb with weak stems to 50 cm long; lvs to 3 cm long excl. the stalk; fls (8a) with petals to 6 mm long, mostly solitary. FR, NL, YP, KI, SE; NSW, Tas.; New Zealand.

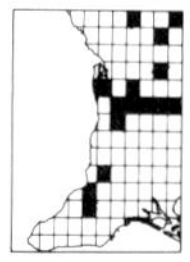

9. *Geranium retrorsum*, **common cranes bill**: Herb with weak stems to 50 cm long; lvs to 3 cm long excl. the stalk; fls (9a) with petals 5–10 mm long, mostly paired; note hairs on fl stalks bent strongly backwards. FR, EP, NL, MU, YP, KI, SE; all States except NT; New Zealand.

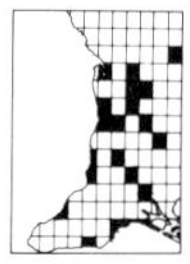

10. *Geranium solanderi*, **native geranium**: Herb with weak stems to 60 cm long; lvs to 3 cm long excl. the stalk; fls (10a) with petals to c. 10 mm long, mostly paired. FR, EA, EP, NL, MU, YP, KI, SE; all States except NT; New Zealand. 5 other species.

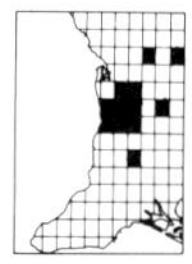

11. *Pelargonium littorale*, a **storks bill**: Herb to 40 cm tall; lvs to 4 cm long excl. the stalk; fls (11a) with petals c. 6 mm long; fertile stamens 3–6. FR, EP, MU, YP, KI, SE; WA, Vic.

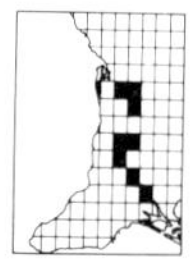

12. *Pelargonium australe*, **austral storks bill**: Perennial herb to 50 cm tall; lvs to 9 cm long excl. the stalk; fls (12a) with petals c. 8 mm long; 6–8 fertile stamens. EP, MU, YP, KI, SE; all States except NT. 2 other introduced species.

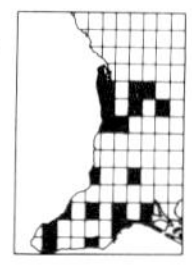

1a
1b
1
2a
2
3
4b
4a
4
5
5a
5b
6a
6b
6
7a
7b
7
8
8a
9a
9
10a
10
11a
11
12a
12

PLATE XL

TROPAEOLACEAE

1. **Tropaeolum majus*, **nasturtium**: Annual herb with stems along the ground or climbing to 1 m; lvs 5–15 cm diam.; fls yellow to red, 2.5–7.0 cm diam. NSW; native to S. America.

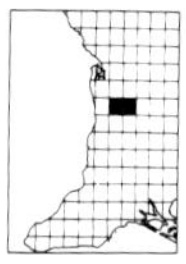

ZYGOPHYLLACEAE

2. *Nitraria billardierei*, **nitre-bush**: Shrub 1–2 m tall; lvs thick, 1–4 cm long; fls (2a), female with petals 3–4 mm long; fruit (2b) 1–2 cm long. ?NW, LE, NU, GT, FR, EA, EP, NL, MU, YP, KI, SE; all mainland States.

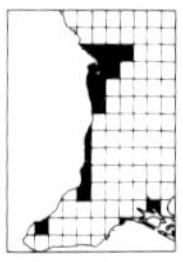

3. *Tribulus terrestris*, **caltrop**: Annual herb with branches along the ground; lvs with 4–7 pairs of 5–10 mm long leaflets; fls with petals 5–6 mm long; fruit (3a) 6–10 mm long. NW, LE, NU, EA, EP, NL, MU, YP; all mainland States; on all continents; possibly native.

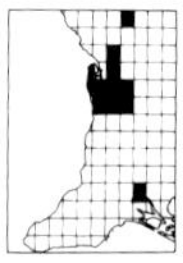

4. *Zygophyllum billardierei*, **coast twinleaf**: Perennial shrub to 50 cm tall, with erect or prostrate branches; lvs 12–20 mm long excl. the stalk; fls with yellow petals 10–12 mm long; fruit (4a) 8–12 mm long, strongly angled but not winged. NU, EP, NL, MU, YP, KI, SE; WA, Vic., Tas. 2 other species.

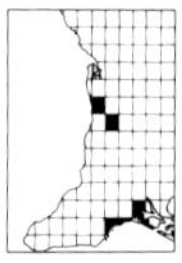

EUPHORBIACEAE

5. *Beyeria lechenaultii*, **pale turpentine bush**: Sticky shrub 50–150 cm tall; lvs usually 9–20 mm long; male fls (5a) 1–2 mm long; female fls (5b) c. 2 mm long; fruit (5c) 3–7 mm long. FR, EA, EP, NL, MU, YP, KI, SE; all States except NT and Qld.

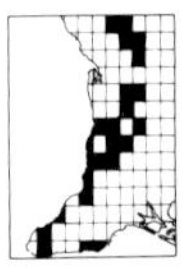

6. *Micrantheum demissum*, **dwarf micrantheum**: Shrub 10–50 cm tall; lvs 3–8 mm long; male fls (6a) 1.5–2.5 mm long; female fls (6b) to 3 mm long; fruit (6c) 6–7 mm long. KI.

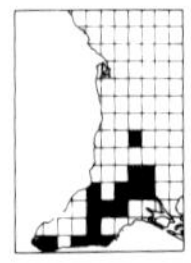

7. **Euphorbia peplus*, **petty spurge**: Erect annual herb to 40 cm tall; lvs 5–25 mm long excl. the stalk; infl (7a) c. 1 mm diam., with central stalked female fl. FR, EP, NL, MU, YP, KI, SE; all States; native to Europe.

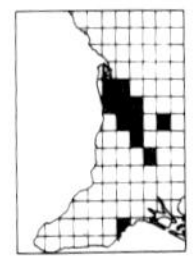

8. *Euphorbia drummondii*, **caustic weed**: Prostrate annual or perennial herb with stems to 30 cm long; lvs 4–8 mm long; infl (8a) c. 1 mm diam., with central stalked female fl; white petal-like structure usually absent or small. All regions except KI; all States. This species and the similar **E. maculata*, **eyebane**, occur in paths and lawns.

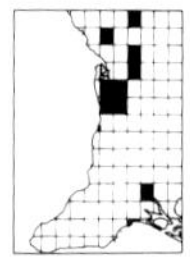

9. **Euphorbia paralias*, **sea spurge**: Perennial fleshy shrub to 70 cm tall; lvs 5–20 mm long; infl (9a) c. 1.5 mm diam., with central stalked female fl; occurs on sandy sea-shore. EP, YP, KI, SE; WA, Vic., Tas.; native to the Mediterranean. 12 or 13 other species.

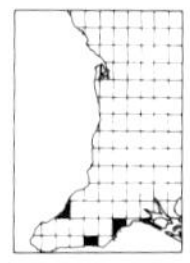

10. *Phyllanthus australis*, **pointed spurge**: Shrub 5–30 cm tall; lvs 3–9 mm long; male (10a) and female (10b) fls with sepals 1.5–2.0 mm long, petals 0; fruit (10c) 3–4 mm diam. ?NW, FR, YP, KI; ?Tas. 3 other species.

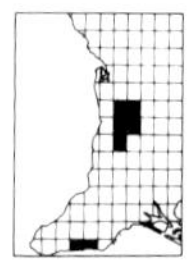

11. *Poranthera ericoides*, a **poranthera**: Shrub 5–25 cm tall; lvs 1–2 cm long; male (11a) and female (11b) fls with sepals c. 2 mm long, petals minute; fruit (11c) 2–3 mm diam. EP, KI; WA.

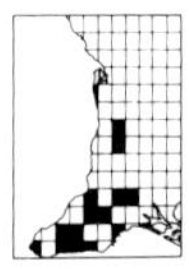

12. *Poranthera microphylla*, **small-leaved poranthera**: Annual herb 2–15 cm tall; lvs usually 5–10 mm long; male (12a) and female (12b) fls with sepals 1.0–1.5 mm long, petals minute; fruit (12c) to 2 mm diam. NU, GT, NL, MU, YP, KI, SE; all States; New Zealand. *P. triandra*, **three-petal poranthera**, is an annual herb with 3 sepals.

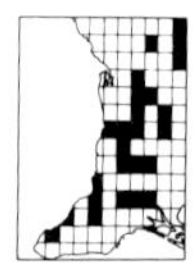

1
2
2a
2b
3
3a
3b
4
4a
5
5a ♂
5b
5c
6
6a
6b
6c
7
7a
8
8a
9
9a
10
10a
10b
10c
11
11a
11b
11c
12
12a
12b
12c

Plate XLI

RUTACEAE

1. *Boronia coerulescens*, **blue boronia**: Shrub to 50 cm tall; lvs usually warty, 1.5–7.0 mm long; fls (1a) with blue, purplish or white petals 3–5 mm long. NU, EP, NL, MU, YP, KI, SE; WA, NSW, Vic.

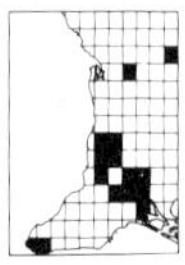

2. *Boronia edwardsii*, **island boronia**: Shrub to 50 cm tall; lvs 2.2–8.0 mm long excl. the stalk; fls (2a) with pink petals 4–7 mm long. MU, KI.

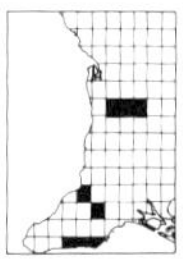

3. *Boronia filifolia*, **slender boronia**: Shrub to 50 cm tall; lvs sometimes compound, 4–20 mm long when simple; fls (3a) with pink petals 4–12 mm long. EP, MU, KI, SE; Vic.

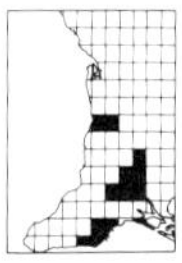

4. *Boronia nana*, **dwarf boronia**: Shrub to 25 cm tall; lvs sometimes not compound, 4–25 mm long excl. the stalk; fls (4a) with white or pale-pink petals 3–4 mm long. SE; NSW, Vic., Tas. *B. inornata*, **desert boronia**, has compound lvs and glandular-warty branchlets.

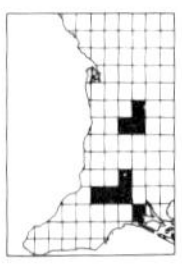

5. *Boronia parviflora*, **swamp boronia**: Shrub to 30 cm tall; lvs 6–20 mm long; fls (5a) with pink or white petals 4–6 mm long. KI, SE; NSW, Vic.

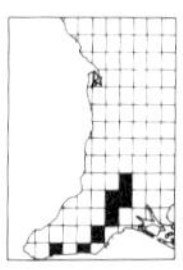

6. *Correa aemula*, **hairy correa**: Shrub to 1 m tall; lvs (6a) 2–4 cm long excl. the stalk; fls (6b) with green or grey-green petals 1.5–3.0 cm long; note calyx lobes. MU, KI; Vic.

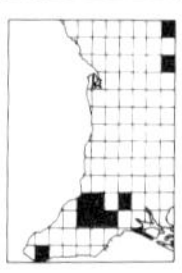

7. *Correa alba*, **white correa**: Shrub 0.5–1.0 m tall; lvs 5–15 mm long excl. the stalk; fls (7a) with white to pink petals 10–15 mm long. SE; Vic.

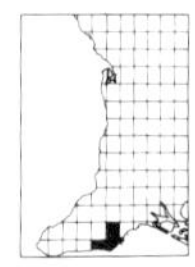

8. *Correa calycina*, **South Australian green correa**: Shrub 1–3 m tall; lvs 2–4 cm long excl. the stalk; fls with green petals 2–3 cm long. KI.

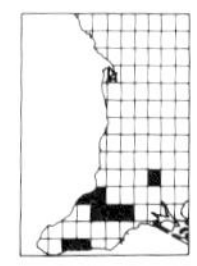

9. *Correa decumbens*, **spreading correa**: Shrub 20–100 cm tall, often with branches along the ground; lvs 2–5 cm long excl. the stalk; fls with pink-red and yellow-green petals 1.5–3.0 cm long; note the calyx lobes with narrower intermediate points. KI.

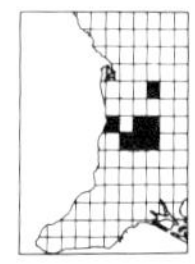

10. *Correa glabra*, **rock correa**: Shrub 1–3 m tall; lvs 1–5 cm long excl. the stalk; fls (10a) with mainly green petals 2–3 cm long. FR, EA, NL, MU, ?KI, SE; QLD, NSW, Vic.

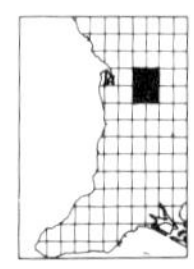

11. *Correa pulchella*, **salmon correa**: Shrub 30–100 cm tall, often with branches along the ground; lvs 1–2 cm long excl. the stalk; fls with pink-red, orange or white petals 1–3 cm long. EP, YP, KI, SE.

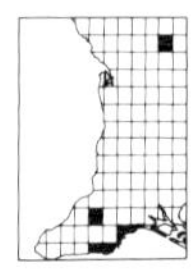

12. *Correa reflexa*, **common correa**: Shrub 50–200 cm tall, sometimes with branches along the ground; lvs 1.0–3.5 cm long excl. the stalk; fls with green to red petals with green or yellow tips, 1.5–4.0 cm long. NU, GT, EP, NL, MU, YP, KI, SE; all States except NT.

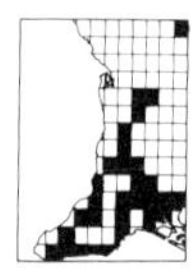

(*Continued*)

1a
1
2a
2
3a
3
4a
4
5a
5
6a
6b
6
7a
7
8
9
10a
10
11
12

RUTACEAE
(Continued)

1. *Eriostemon angustifolius*, **narrow-leaved wax-flower**: Shrub to 60 cm tall; lvs with small black stipules, 2–10 mm long; fls with 5 white to pink petals 5–9 mm long. FR, EP, NL, MU, YP, KI, SE; Qld, Vic. *E. pungens*, **prickly wax-flower**, has pointed lvs.

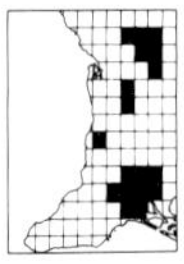

2. *Phebalium hillebrandii*, **Hillebrands phebalium**: Shrub to 60 cm tall; lvs 3–17 mm long; fls with 5 white to pink petals 3.5–5.0 mm long. KI, SE. 2 other species.

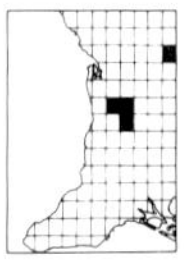

3. *Zieria veronicea*, **pink zieria**: Shrub to 60 cm tall; lvs 5–15 mm long; fls with 4 pink or white petals 2.3–7.0 mm long. MU, KI, SE; Vic., Tas.

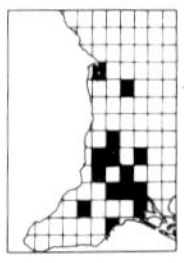

LINACEAE

4. *Linum marginale*, **native flax**: Annual or perennial herb 10–50 cm tall; lvs 5–20 mm long; fls 6–12 mm long. FR, EA, EP, NL, MU, YP, SE; all States except NT; New Zealand.

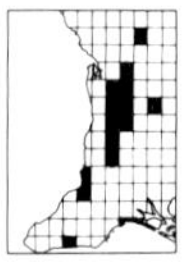

5. **Linum trigynum*, **French flax**: Annual herb 10–50 cm tall; lvs 5–20 mm long; fls 5–7 mm long. NL, YP, KI; all States except NT and Tas.; native to Europe and the Mediteranean. 2 other species.

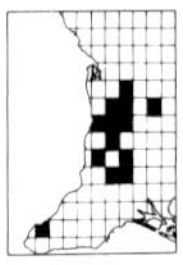

TREMANDRACEAE

6. *Tetratheca pilosa*, **pink-eyed Susan**: Shrub to 60 cm tall; lvs 2–15 mm long; fls white to dark-pink, with the 3–12 mm long petals spreading flat or forming a cup shape. MU, YP, SE; NSW, Vic., Tas.

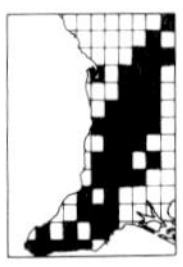

POLYGALACEAE

7. *Comesperma calymega*, **blue-spike milkwort**: Shrub 10–70 cm tall; lvs 4–25 mm long; fls (7a) with wings 4–5 mm long; fruit (7b) 7–9 mm long, enclosing long-haired seeds. EP, MU, KI, SE; WA, Vic., Tas. *C. polygaloides*, **small milkwort**, is usually a smaller plant with larger wings.

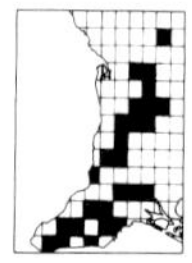

8. *Comesperma volubile*, **love creeper**: Climbing perennial 20–200 cm tall; lvs 0–8 cm long; fls (8a) with wings c. 6.5 mm long. NU, EP, NL, MU, YP, KI, SE; all States except NT.

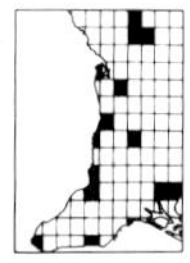

9. **Muraltia heisteria*, **furze muraltia**: Prickly shrub to 1 m tall; lvs 5–12 mm long; fls (9a) sometimes white, with wings 4–5 mm long. NSW, Vic.; native to S. Africa.

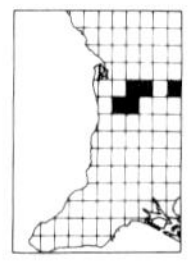

SAPINDACEAE

10. *Dodonaea baueri*, **crinkled hop-bush**: Shrub to 1 m tall; lvs 5–18 mm long; fruit lobed but not winged, 4–5 mm long. FR, EA, EP, NL, MU, YP, KI.

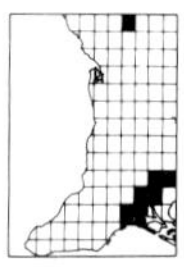

11. *Dodonaea viscosa*, **sticky hop-bush**: Shrub 1–4 m tall; lvs (11a and 11b variation) 1–10 cm long; fls (11a male) with sepals 1.3–3.0 mm long, petals 0. All regions; all States; worldwide. 4 other species.

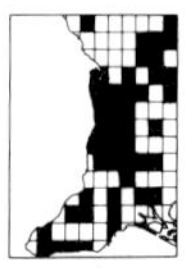

STACKHOUSIACEAE

12. *Stackhousia aspericocca*, **candles**: Perennial herb to 70 cm tall; lvs to c. 35 mm long; fls (12a) white to deep-yellow, 8–14 mm long; note small conspicuous bract on either side of fl base. EP, NL, MU, YP, KI, SE; Vic.

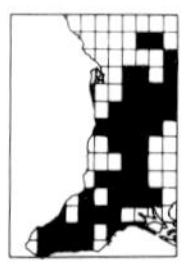

13. *Stackhousia monogyna*, **creamy candles**: Perennial herb to 50 cm tall; lvs to 35 mm long; fls (13a) cream-white to light-yellow, 8–14 mm long; note the swollen bract base. FR, EP, NL, MU, YP, KI, SE; ?Qld, NSW, Vic., Tas.

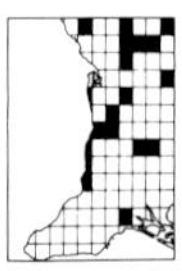

1
2
3
4
5a
5
6
7a
8a
8
7
7b
9a
9
11b
10
11a
11
11c
12a
12
13a
13

Plate XLIII

RHAMNACEAE

1. *Cryptandra hispidula*, **rough cryptandra**: Shrub 20–30 cm tall; lvs 3–6 mm long; fls (1a) 4–6 mm long, in dense heads; petals tiny, hooding the anthers. KI.

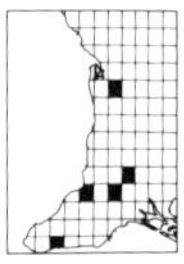

2. *Cryptandra leucophracta*, **rusty poison**: Shrub to 10 cm or more tall; lvs 4–6 mm long; fls (2a) 4–6 mm long; in dense heads with white leafy bracts; petals tiny. FR, EP, NL, MU, YP, KI, SE; WA, NSW, Vic.

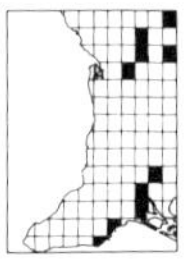

3. *Cryptandra tomentosa*, **prickly cryptandra**: Shrub 25–30 cm tall; lvs 2–6 mm long; fls (3a) c. 3 mm long in loose infls; petals tiny. EP, NL, MU, YP, KI, SE; WA, Vic.

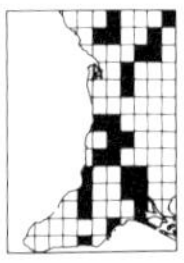

4. *Pomaderris oraria*, **coast pomaderris**: Shrub 30–50 cm tall; lvs 8–25 mm long; fls (4a) with sepals c. 2 mm long, petals 0. FR, EA, EP, NL, MU, YP, KI, SE; all States except NT and Qld. *P. obcordata*, **wedge-leaved pomaderris**, has notched or 2-lobed lvs.

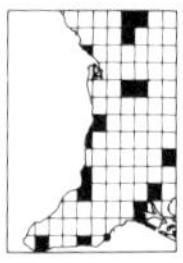

5. *Rhamnus alaternus*, **blow-fly bush**: Shrub or tree to 5 m tall; lf blades 2.0–7.5 cm long; fls (5a) with sepals c. 2 mm long, petals 0; fruit (5b) c. 5 mm long. EP, NL, MU, YP, SE; NSW, Vic.; native to the Mediterranean.

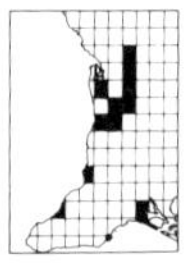

6. *Spyridium coactilifolium*, **butterfly spyridium**: Shrub 30–200 cm tall; lvs 6–15 mm long; fl head (6a); fls c. 3 mm long; petals minute, hooding the anthers. SE.

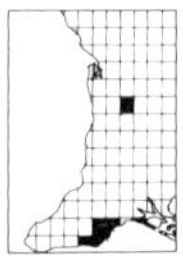

7. *Spyridium parvifolium*, **dusty miller**: Shrub 1–2 m tall; lvs 5–15 mm long; fls (7a) 1.5–2.0 mm long. FR, EP, NL, MU, ?KI, SE; NSW, Vic., Tas.

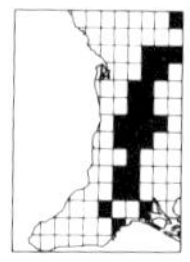

8. *Spyridium eriocephalum*, **heath spyridium**: Shrub 30–150 cm tall; lvs 3–10 mm long; fl head (8a); fls c. 2 mm long. EP, NL, MU, YP, KI, SE; Vic., Tas.

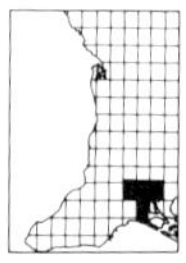

9. *Spyridium spathulatum*, **spoon-leaved spyridium**: Shrub 1–2 m tall; lvs 5–15 mm long; fls (9a) c. 1.5 mm long. FR, EP, MU, KI, SE; WA, Vic.

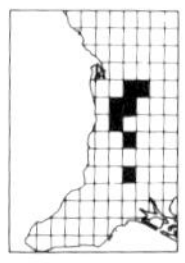

10. *Spyridium thymifolium*, **thyme-leaved spyridium**: Shrub to 60 cm tall; lvs 7–20 mm long; fls (10a) c. 2 mm long. KI, SE.

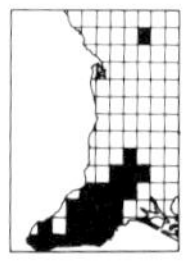

11. *Spyridium vexilliferum*, **winged spyridium**: Shrub to 60 cm tall; lvs 8–15 mm long; fl head (11a) 4–6 mm diam.; fls (11b) c. 2 mm long. EP, NL, KI, SE; Vic., Tas. *S. phylicoides*, narrow-leaved spyridium, but narrower lvs. *S. subochreatum*, **velvet spyridium**, has conspicuous stipules at the base of the lvs and dense short star-shaped hairs on the fls and lvs.

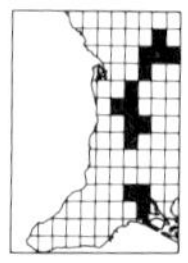

12. *Trymalium wayae*, **grey trymalium**: Shrub 1–2 m tall; lvs 5–25 mm long excl. the short stalk; fls (12a) with sepals c. 1 mm long; petals 0. ?EP, MU, KI, SE.

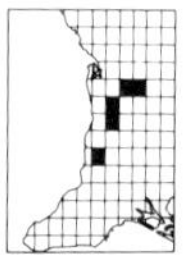

1a
1
2
2a
3a
3
4
4a
5
5a
5b
6a
6
7a
7
8
8a
9a
9
10a
10
11a
11b
11
12a
12

Plate XLIV

MALVACEAE

1. *Hibiscus trionum*, **bladder ketmia**: Annual to c. 60 cm tall; lvs variously lobed, the blades 15-60 mm long; fls (1a) pale-yellow with purplish centre, petals 15-27 mm long; fruit 10-16 mm long. LE, FR, EP, NL, MU, SE; all States; probably native to Europe and Asia.

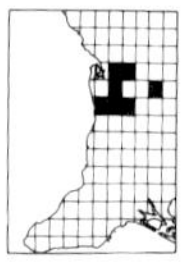

2. *Lavatera plebeia*, **Australian hollyhock:** Perennial herb to 4 m tall; lf blades 1-20 cm long; fls (2a and 2b) white to lilac, with petals 20-30 mm long; fruit 2.0-3.5 mm long. All regions; all States. **L. cretica*, Cretan hollyhock, has lilac fls, the calyx not longer than the fruit, which it is in *L. plebeia*.

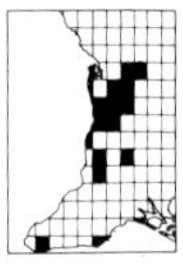

3. **Lavatera arborea*, **tree mallow**: Biennial 1-3 m tall; lf blades 2-12 cm long; fls (3a and 3b) with petals 15-20 mm long; fruit 2.5-4.0 mm long. EP, NL, MU, YP, KI, SE; WA, NSW, Vic.; New Zealand; native to Europe and the Mediterranean.

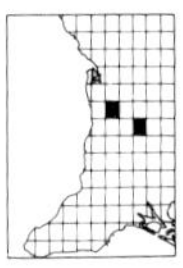

4. **Alcea rosea*, **hollyhock**: Biennial or perennial herb to 2.5 m tall; lf blades 4-14 cm long; fls (4a) with white or yellow to dark-red or purple petals c. 4.5 cm long; fruit (4b) c. 17 mm long. NL, MU, ?YP; NSW; probably of hybrid origin.

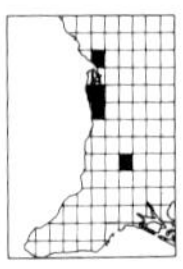

5. *Lawrencia squamata*, **thorny lawrencia**: Shrub often 50-100 cm tall; lvs (5a) 3-10 mm long; fls (5b) with petals 3-5 mm long; fruit c. 2 mm long. LE, NU, GT, FR, EA, EP, NL, MU, YP, SE; all mainland States. 2 other species.

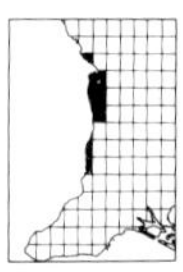

6. **Malva nicaeensis*, **mallow of Nice**: Annual or perennial herb to c. 50 cm tall; lf blades 1-7 cm long; fls (6a and 6b) with petals 4-12 mm long; fruit (6c segment of fruit) 5-7 mm diam. NL, KI, SE; NSW, Vic., Tas.; native to Europe.

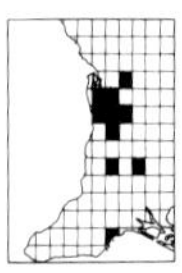

7. **Malva parviflora*, **small-flowered marshmallow**: Annual or perennial herb to 150 cm tall; lf blades 0.6-9.0 cm long; fls (7a) with petals 3-5 mm long; fruit (7b 1 segment of fruit) 5.5-8.0 mm diam. All regions; all States; native to the Mediterranean.

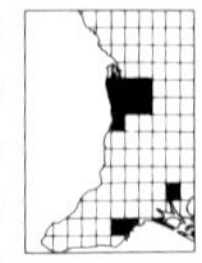

8. **Modiola caroliniana*, **red-flowered mallow**: Perennial herb usually under 25 cm tall; lf blades 1-4 cm long; fls (8a) with petals 4-6 mm long; fruit (8b 1 segment of fruit) c. 8 mm diam. SE; all States except NT; New Zealand; native to S. America.

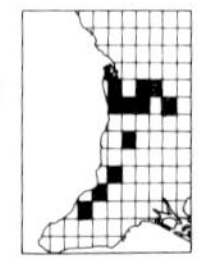

9. *Sida corrugata*, **corrugated sida**: Low spreading perennial more or less prostrate to 10 cm tall; lf blades 5-25 mm long; fls with petals 5-6 mm long; fruit (9a 1 segment of fruit) 4.0-5.5 mm diam. ?NW, LE, NU, GT, FR, EA, EP, NL, MU, YP, SE; WA, ?Qld, NSW, Vic. *S. petrophila*, **tall sida**, is a sparsely branched erect shrub to 2 m tall with larger fls.

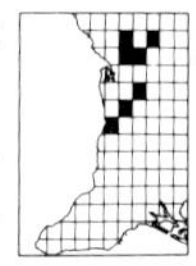

VIOLACEAE

10. *Viola sieberana*, **tiny violet**: Perennial herb with stolons; lvs 15-55 mm long incl. stalk; fls (10a) with petals 5-7 mm long; fruit 4-5 mm long. KI, SE; Vic. 4 other species.

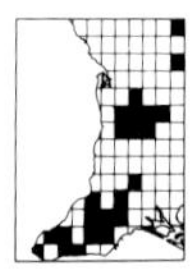

11. *Hybanthus floribundus*, **shrub violet**: Shrub to 1.5 m tall; lvs 5-35 mm long; fls (11a and 11b) with the largest petals 4-11 mm long; fruit to 8 mm long. FR, EP, NL, MU, YP, KI, SE; WA, NSW, Vic. *H. monopetalus*, **slender violet-bush**, has green sepals.

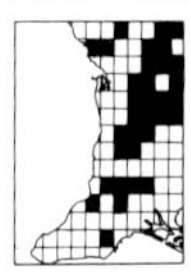

12. *Hymenanthera dentata*, **tree violet:** Shrub to 2 m tall; lvs 5-20 mm or more long; fls (12a) with petals 3-4 mm long; fruit a berry, 4-5 mm diam. FR, NL, MU, SE; NSW, Vic., Tas.; New Zealand.

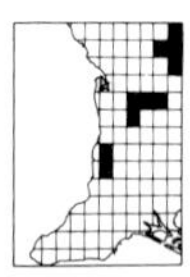

1a
1
2a
2b
2
3b
3a
3
4b
4a
4
5b
5a
5
6a
6b
6c
6
7c
7a
7
7b
8a
8
8b
9a
9
10a
10
11a
11b
11
12a
12

Plate XLV

STERCULIACEAE

1. *Lasiopetalum baueri*, **slender velvet-bush**: Shrub to 1.5 m tall; lvs 2–5 cm long; fls (1a and 1b) pink or white, with petals sepals-like 4–6 mm long; fruit 4–8 mm diam. EP, MU, YP, KI, SE; NSW, Vic., Tas. 2 other species.

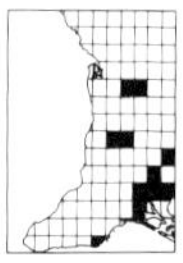

2. *Thomasia petalocalyx*, **paper-flower**: Shrub to 1 m tall; lvs 2.0–4.5 cm long; fls mauve, with petal-like 5–10 mm long; fruit 7–9 mm long; note the large paired leaf-like stipules at the base of the lvs. EP, NL, MU, YP, KI, SE; WA, Vic.

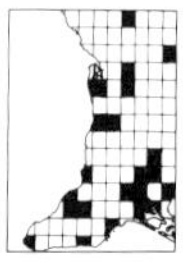

THYMELAEACEAE

3. *Pimelea curviflora*, **curved riceflower**: Shrub to 1.5 m tall; lvs 2–20 mm long; fls (3a) greenish-yellow and often red on the lobes, 3.0–10.5 mm long. FR, NU, EP, NL, MU, YP, KI, SE; all States except WA and NT.

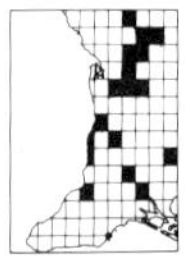

4. *Pimelea flava*, **yellow riceflower**: Shrub to 2 m tall; lvs 2–10 mm long; fls (4a female, 4b male) yellow or white, 3.5–8.5 mm long. FR, EP, MU, YP, KI, SE; NSW, Vic., Tas.

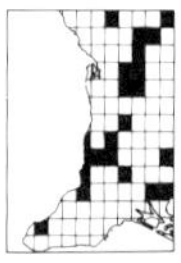

5. *Pimelea glauca*, **smooth riceflower**: Shrub to 1 m tall; lvs 3–20 mm long; fls (5a) creamy-white, 10–18 mm long. FR, EP, NL, MU, YP, KI, SE; all States except WA and NT.

6. *Pimelea humilis*, **common riceflower**: Shrub to 0.5 m tall; lvs 5–20 mm long; fls (6a) creamy-white, 13–18 mm long. FR, EP, NL, MU, SE; NSW, Vic., Tas.

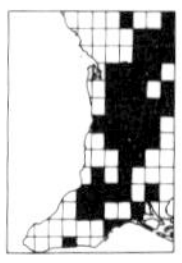

7. *Pimelea linifolia*, **slender riceflower**: Shrub to 1.5 m tall; lvs 3–32 mm long; fls (7a) white, 10–20 mm long. FR, EP, SE; all States except WA and NT.

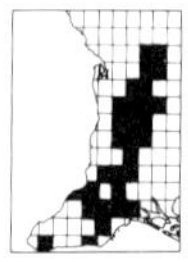

8. *Pimelea octophylla*, **woolly riceflower**: Shrub to 1 m tall; lvs 2–15 mm long; fls (8a) cream to pale-yellow, 6.5–18.5 mm long. LE, FR, EP, NL, MU, YP, KI, SE; WA, Vic.

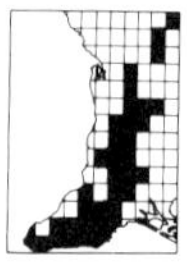

9. *Pimelea phylicoides*, **heath riceflower**: Shrub to 1 m tall; lvs 2.0–8.5 mm long; fls (9a) white, 7.0–13.5 mm long. EP, MU, KI, SE; Vic.

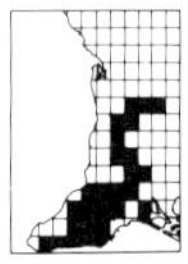

10. *Pimelea serpyllifolia*, **thyme riceflower**: Shrub to 1.5 m tall; lvs 2–13 mm long; fls (10a male, 10b female) yellow, 3–4 mm long. NU, EP, NL, MU, YP, KI, SE; WA, NSW, Vic.

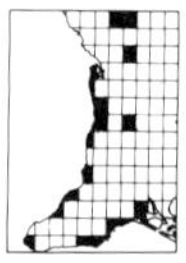

11. *Pimelea stricta*, **gaunt riceflower**: Shrub to 1.5 m tall; lvs 5.5–37.0 mm long; fls (11a) creamy-white, 10–16 mm long. FR, EA, EP, NL, MU, KI, SE; NSW, Vic. *P. microcephala*, a **riceflower**, has hairless stems and greenish-yellow slightly hairy fls. *P. simplex*, **desert riceflower**, is herbaceous and its fls are in elongating spikes.

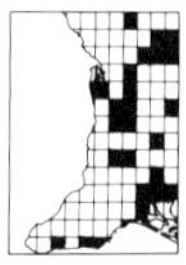

FRANKENIACEAE

12. *Frankenia pauciflora*, **common sea-heath**: Shrub usually under 30 cm tall; lvs 3–13 mm long; fls (12a and 12b) 6.5–10.5 mm long when stretched out. LE, EP, NL, MU, YP, KI, SE; Vic., Tas. *F. pulverulenta*, a **sea-heath**, has smaller fls.

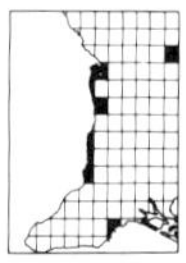

1b
1a
1
2
3a
3
4b
♂
4a
♀
4
5
5a
6
6a
7
7a
8a
8
9a
9
10a
♂
10b
♀
10
11
11a
12a
12b
12

Plate XLVI

MYRTACEAE

1. *Kunzea pomifera*, **muntries**: Shrub low on the ground, to 2 m or more in diam.; lvs 4–8 mm long; fls 5–10 mm long; fruit (1a) fleshy, 8–12 mm diam.; edible. MU, YP, KI, SE; Vic.

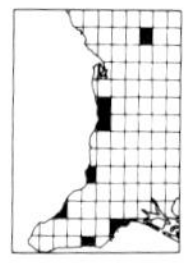

2. *Baeckea ramosissima*, **rosy baeckea**: Shrub to 60 cm tall; lvs 3–8 mm long; fls (2a and 2b) white to mauve-purple, 3–15 mm across; fruit (2c) c. 2 mm long. KI, SE; NSW, Vic., Tas. 2 other species.

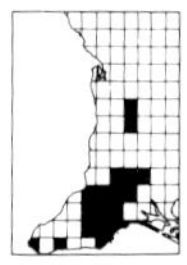

3. *Calytrix tetragona*, **common fringe-myrtle**: Shrub to 3 m tall; lvs 0.75–14.00 mm long; fls (3a and 3b) with petals 3.5–7.5 mm long, the calyx (3c) with awns to 17 mm long. FR, EA, EP, NL, MU, YP, KI, SE; all States except NT. *C. glaberrima*, **smooth fringe-myrtle**, lacks the calyx awns.

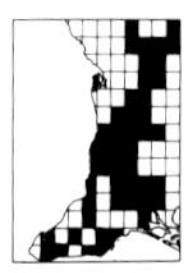

4. *Leptospermum sp.*, **prickly tea-tree**: Shrub 1–2 m or more tall; lvs 5–15 mm long; fls (4a) c. 10 mm diam.; fruit (4b) 6–7 mm diam. Formerly known as *L. juniperinum*. NL, MU, KI, SE; NSW, Vic.

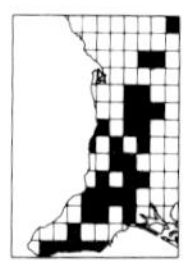

5. *Leptospermum lanigerum*, **silky tea-tree:** Shrub to 5 m or more tall; lvs to 15 mm long; fls (5a and 5b) c. 15 mm diam.; fruit 5–10 mm diam. NL, MU, KI, SE; NSW, Vic., Tas.

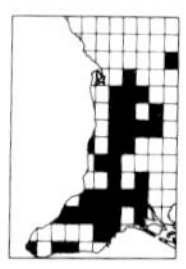

6. *Leptospermum myrsinoides*, **heath tea-tree**: Shrub usually 1–2 m tall; lvs 5–10 mm long; fls 10–15 mm diam.; fruit (6a) 4–6 mm diam. NL, MU, KI, SE; NSW, Vic. *L. laevigatum*, **coastal tea-tree**, may have been introduced from the eastern States, having larger lvs (15–30 mm long) and fls (15–20 mm diam.).

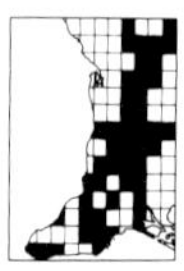

7. *Eucalyptus baxteri*, **brown stringybark**: Usually single-stemmed tree to 30 m tall; bark fibrous and furrowed; lvs (7a) 3–8 cm long; buds (7b) 6–15 mm long; fruit (7c) 8–12 mm long. MU, KI, SE; Vic.

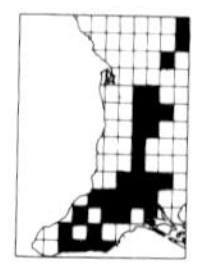

8. *Eucalyptus goniocalyx*, **long-leaved box**: Multi-stemmed or single-stemmed tree to 15 m tall; bark rough; lvs 10–27 cm long; juvenile lvs (8a); buds 9–14 mm long; fruit (8b) 8–9 mm long. FR, NL; NSW, Vic.

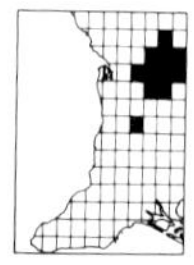

9. *Eucalyptus microcarpa*, **grey box**: Multi-stemmed or single-stemmed tree to 25 m tall; bark rough; lvs 7–14 cm long; buds 6–7 mm long; fruit (9a) 4–5 mm long. FR, NL, SE; Qld, NSW, Vic.

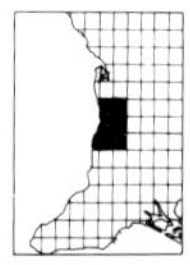

10. *Eucalyptus obliqua*, **messmate stringybark**: Usually single-stemmed tree to 30 m tall; bark fibrous; lvs 8–16 cm long; buds 6–8 mm long; fruit (10a) 7–12 mm long. KI, SE; NSW, Vic.

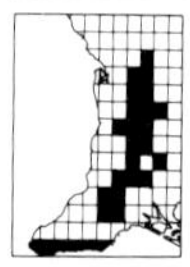

11. *Eucalyptus odorata*, **peppermint box**: Multi-stemmed or single-stemmed tree to 20 m tall; bark rough on the trunk; lvs dull, green to grey-green, 5–14 cm long; buds 4–7 mm long; fruit (11a) 5–8 mm long. FR, EP, NL, MU, YP, KI, SE; Vic.

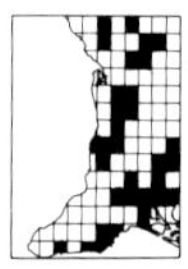

12. *Eucalyptus porosa*, **mallee box**: Multi-stemmed or single-stemmed tree to 10 m tall; bark rough; lvs glossy, green to dark-green, 6–9 cm long; buds 5–7 mm long; fruit (12a) 5–8 mm long. GT, FR, EA, EP, NL, MU, YP, SE; NSW, Vic.

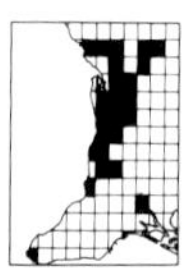

(*Continued*)

1a
1
2a
2b
2c
2
3a
3b
3c
3
4a
4b
4
5a
5b
5c
5
6a
6
7c
7b
7a
8
8a
8b
9a
9
10
10a
11
11a
12a
12

Plate XLVII

MYRTACEAE

(*Continued*)

1. *Eucalyptus calycogona*, **square-fruited mallee**: Multi-stemmed or single-stemmed tree to c. 10 m tall; bark smooth; lf blades 5–10 cm long; buds 8–14 mm long; fruit (1a) 8–12 mm long. FR, EP, NL, MU, YP, SE; WA, NSW, Vic.

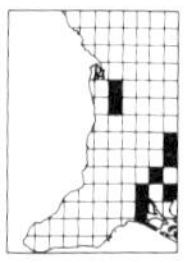

2. *Eucalyptus camaldulensis*, **river red gum**: Single-stemmed tree usually less than 20 m tall; bark smooth, often rough near ground; lf blades 6–20 cm long; buds 6–12 mm long; fruit (2a) 5–8 mm long. NW, LE, FR, EA, EP, NL, MU, YP, KI, SE; all mainland States.

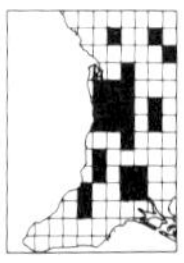

3. *Eucalyptus cneorifolia*, **Kangaroo Island narrow-leaved mallee**: Multi-stemmed or single-stemmed tree to 10 m tall; bark rough below; lf blades 5–10 cm long; buds 7–10 mm long; fruit (3a) 5–7 mm long. KI.

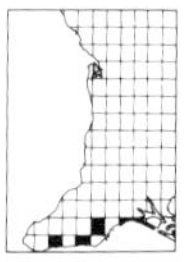

4. *Eucalyptus cosmophylla*, **cup gum**: Single-stemmed or multi-stemmed tree to 5 m tall; bark smooth; lf blades 10–15 cm long; buds in threes, 12–15 mm long; fruit (4a) 10–17 mm long. KI.

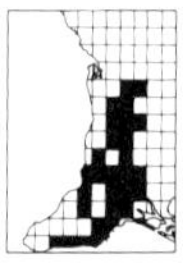

5. *Eucalyptus diversifolia*, **coastal white mallee gum**: Multi-stemmed tree to 4 m tall; bark smooth; lf blades usually 6–10 cm long; buds 7–10 mm long; fruit 7–12 mm long. NU, EP, YP, KI, SE; WA, Vic.

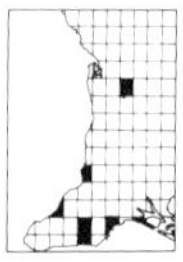

6. *Eucalyptus fasciculosa*, **pink gum**: Single-stemmed or multi-stemmed tree to 15 m tall; bark smooth; lf blades 8–16 cm long; buds (6a) in branched infls, 4–6 mm long; fruit (6b) 6–10 mm long. MU, KI, SE; Vic.

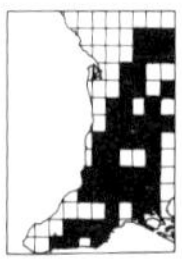

7. *Eucalyptus leptophylla*, **narrow-leaved mallee**: Multi-stemmed or single-stemmed tree to 9 m tall; bark smooth; lf blades 3–10 cm long, glossy; buds 6–8 mm long; fruit (7a and 7b) 3–6 mm long. NW, NU, GT, FR, EA, EP, NL, MU, YP, KI, SE; WA, NSW, Vic.

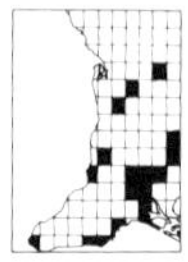

8. *Eucalyptus leucoxylon*, **South Australian blue gum**: Usually single-stemmed tree to 30 m tall; bark smooth, often rough towards ground; lf blades 7–25 cm long; buds (8a) in threes, 7–20 mm long; fls white to yellow or red; fruit (8b) to 25 mm long. FR, EP, NL, MU, KI, SE; Vic.

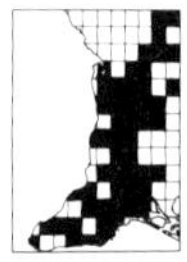

9. *Eucalyptus oleosa*, a **red mallee**: Usually multi-stemmed tree to 10 m tall; bark mainly smooth; lf blades 6–14 cm long; buds 6–8 mm long; fruit (9a) 5–8 mm long, with fragile valves. NW, LE, NU, GT, FR, EA, EP, NL, MU, YP, KI, SE; WA, NSW, Vic.

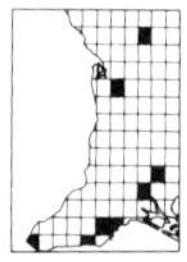

10. *Eucalyptus ovata*, **swamp gum**: Single-stemmed tree to 17 m tall; bark smooth; lf blade 7–14 cm long; buds 7–10 mm long; fruit (10a) 4–8 mm long. KI, SE; NSW, Vic., Tas.

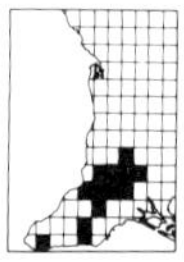

11. *Eucalyptus socialis*, a **red mallee:** Usually multi-stemmed tree to 10 m or more tall; bark mainly smooth; lf blades 6–14 cm long; buds 8–13 mm long, often with a waxy bloom; fruit (11a) 5–10 mm long. NW, LE, NU, GT, FR, EA, EP, NL, MU, YP, KI, SE; WA, NT, NSW, Vic.

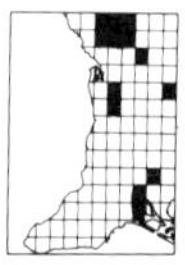

12. *Eucalyptus viminalis*, **manna gum**: Single-stemmed tree to 35 m tall; bark smooth or rough; buds 7–8 mm long; fruit (12a) 6–8 mm long. EP, MU, KI, SE; all States except WA and NT. 6 other species.

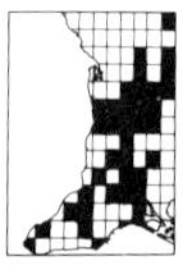

(*Continued*)

1a
1
2
2a
3a
3
4
4a
5
5a
6
6a
7
7a
7b
8
8a
9
9a
10
10a
11
11a
12
12a

Plate XLVIII

MYRTACEAE
(*Continued*)

1. *Callistemon rugulosus*, **scarlet bottlebrush**: Shrub usually under 4 m tall; lvs usually 3–5 cm long; fls (1a) usually 19–24 mm long; fruit usually 7–8 mm long. FR, EP, NL, MU, YP, KI, SE; NSW, Vic.

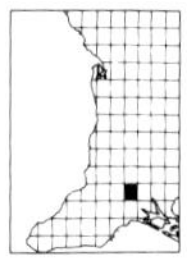

2. *Callistemon sieberi*, **river bottlebrush**: Shrub or tree 2–3 m tall; lvs usually 3–5 cm long; fls 9.0–11.5 mm long; fruit (2a) 3–5 mm long. NL; Qld, NSW, Vic., ?Tas. 2 other species.

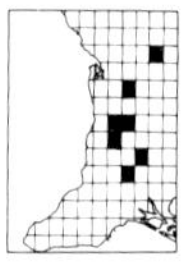

3. *Melaleuca brevifolia*, **mallee honey-myrtle**: Shrub 1–3 m tall; lvs (3a) alternate, 4–8 mm long; fls 6–8 mm long; fruit 3–4 mm long. EP, MU, KI, SE; WA, Vic.

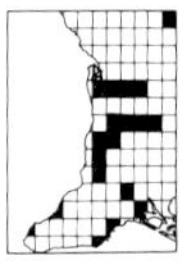

4. *Melaleuca decussata*, **totem poles**: Shrub 1–3 m tall; lvs (4a) opposite, 5–15 mm long; fls 4–6 mm long, light pink to purple; fruit (4b) c. 4 mm long. EP, YP, KI, SE; Vic.

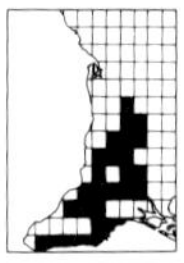

5. *Melaleuca halmaturorum*, **South Australian swamp paper-bark**: Shrub or tree 2–7 m tall; bark papery, white; lvs (5a) opposite, 3–7 mm long; fls 7–9 mm long; fruit 3–4 mm long. EP, MU, YP, KI, SE; Vic.

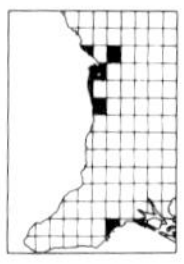

6. *Melaleuca lanceolata*, **moonah**: Shrub or tree to 10 m tall; bark rough, dark; lvs (6a) alternate, 5–15 mm long; fls (6b) 7–10 mm long. NU, GT, FR, EA, EP, NL, MU, YP, KI, SE; WA, Qld, NSW, Vic. *M. pauperiflora*, boree, has smaller fls and usually blunter lvs.

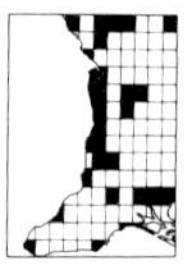

7. *Melaleuca squamea*, **swamp honey-myrtle**: Shrub or tree 1–6 m tall; lvs (7a) alternate 4–8 mm long; fls 8–10 mm long; fruit (7b) 5–6 mm long. KI, SE; NSW, Vic., Tas.

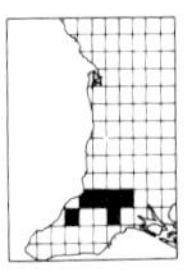

8. *Melaleuca uncinata*, **broombush**: Shrub 1–3 m tall; lvs (8a) alternate, 15–60 mm long; fls 5.0–5.5 mm long; fruit (8b) 1.5–2.5 mm long. NW, LE, NU, GT, FR, EP, NL, MU, YP, KI, SE; all States except NT and Tas.

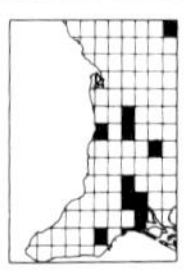

LYTHRACEAE

9. *Lythrum hyssopifolia*, **lesser loosestrife**: Weak annual herb to 50 cm tall; lvs to 25 mm long; fls (9a) with petals 1.5–4.0 mm long; fruit (9b) 4–7 mm long. NW, LE, GT, EA, EP, NL, MU, YP, KI, SE; all States and overseas. Probably introduced. 2 other species with larger fls.

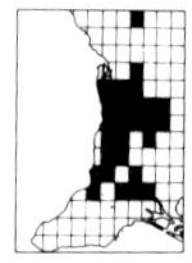

ONAGRACEAE

10. **Oenothera stricta*, **common evening-primrose**: Biennial herb 20–80 cm tall; lvs 3–20 cm long; fls (10a) with petals 2.6–3.0 cm long; fruit (10b) 15–35 mm long; mainly closed during the day. EP, NL, MU, YP, KI, SE; NSW, Vic.; native to America. 4 other species.

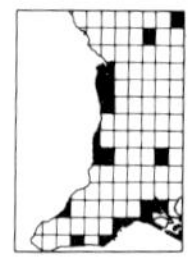

11. *Epilobium billardieranum*, **variable willow-herb**: Perennial herb to 80 cm tall; lvs 1–6 cm long; fls (11a) with petals 3–13 mm long, the ovary with glandular and plain hairs (11b). FR, EP, MU, YP, KI, SE; all States except NT; New Zealand. **E. ciliatum*, a **willow-herb**, has similar lvs but more tapered to the point and is often an annual.

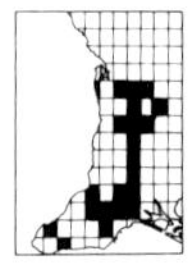

12. *Epilobium hirtigerum*, **hairy willow-herb**: Perennial herb to 1.2 m tall; lvs 1.5–6.5 cm long; fls (12a) with petals white to rose-purple, 3–7 mm long, the ovary without glandular hairs. FR, NL, MU, SE; all States except WA and NT; New Zealand, Java, S. America.

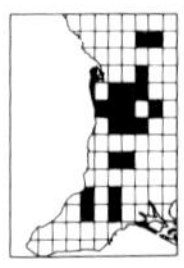

13. *Epilobium pallidiflorum*, **showy willow-herb**: Perennial herb usually less than 70 cm tall; lvs 3–7 cm long; fls (13a) with white or pinkish petals 7–12 mm long, the ovary with glandular and plain hairs. MU, SE; Vic., Tas.; New Zealand.

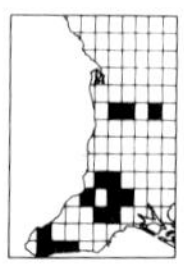

1a
1
2
2a
3
3a
4
4a
4b
5
5a
6
6a
6b
7
7a
7b
8
8a
8b
9
9a
9b
10
10a
10b
11
11a
11b
12
12a
13
13a

Plate XLIX

HALORAGACEAE

1. *Gonocarpus elatus*, **hill raspwort**: Herb or shrub usually 18–35 cm tall; lvs usually 2–3 cm long; fls (1a) 2.1–2.7 mm long excl. ovary; note lvs alternate throughout. FR, EP, NL, MU, YP; Qld, NSW, Vic.

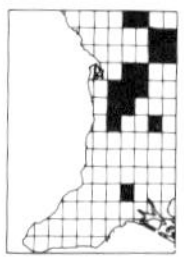

2. *Gonocarpus mezianus*, a **raspwort**: Perennial herb usually 20–30 cm tall; lvs usually 0.7–1.7 cm long, opposite below; fls (2a) 2–3 mm long excl. ovary. EP, NL, YP, KI, SE; Vic. *G. micranthus*, **creeping raspwort**, is a weak plant without hairs.

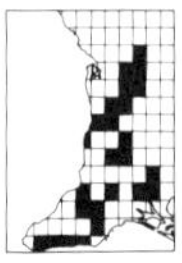

3. *Gonocarpus tetragynus*, **common raspwort**: Perennial herb 15–30 cm tall; lvs usually 0.4–1.2 cm long; fls (3a) 2.0–2.8 mm long excl. ovary; fruit (3b young) 1.0–1.5 mm long. FR, EP, NL, MU, KI, SE; all States except WA and NT.

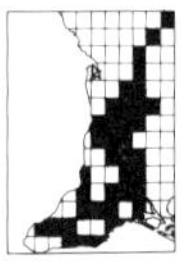

4. *Haloragis acutangula*, **smooth raspwort**: Herb or shrub 20–45 cm tall; lvs usually 1–3 cm long; fls 2.4–3.0 mm long excl. ovary; fruit (4a, 4b and 4c, 3 forms) 1.8–2.5 mm long; note absence of hairs from all parts. NU, EP, NL, MU, YP, KI, SE; WA.

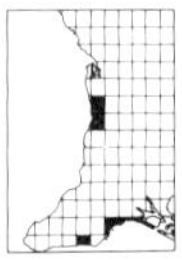

5. *Haloragis aspera*, **rough raspwort**: Perennial herb 12–25 cm tall; lvs 2–5 cm long; fls 2.1–3.0 mm long excl. ovary; fruit (5a) 2.0–4.5 mm long. NW, LE, GT, FR, EA, EP, NL, MU, YP, KI, SE; all States except WA. 3 other species.

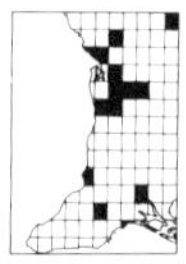

6. *Myriophyllum amphibium*, **broad water-milfoil**: Mat-forming marsh plant 5–7 cm tall; lvs usually 4–10 mm long; fls 1.8–3.0 mm long excl. ovary; fruit (6a) c. 1 mm long. KI; Vic., Tas. 7 other species often having finely divided lvs (6b). *Haloragis brownii*, a **raspwort**, is a similar water plant also with finely divided lvs.

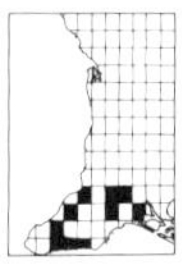

PRIMULACEAE

7. **Anagallis arvensis*, **scarlet** or **blue pimpernel**: Annual herb with weak branches 3–50 cm long; lvs 4–20 mm long; fls red (7a) or blue (7b), 4–15 mm across; fruit 4–6 mm long. LE, NU, GT, FR, EA, EP, NL, MU, YP, KI, SE; all States; native to Europe. **A. minima*, **chaffweed**, has smaller parts and alternate lvs.

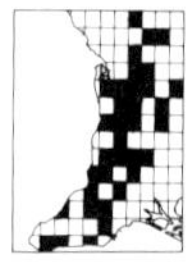

8. *Samolus repens*, **creeping brookweed**: Perennial herb to 60 cm tall; lvs 1–5 cm long; fls (8a) with white or pinkish petals 2–4 mm long. LE, NU, FR, EA, EP, NL, MU, YP, KI, SE; all States except NT.

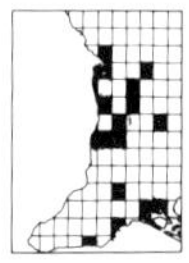

LIMONIACEAE

9. **Limonium companyonis*, a **sea-lavender**: Perennial herb 10–35 cm tall; lvs 1.5–5.0 cm long; fls c. 5 mm long when stretched out. EP, NL, MU, YP, KI, SE; WA; native to Europe. 4 other species.

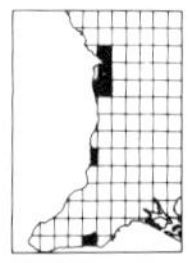

GENTIANACEAE

10. **Centaurium erythraea*, **common centaury**: Annual or biennial herb to 35 cm tall; lvs 1.5–6.0 cm long; fls (10a) 11–17 mm long when stretched out; note the shorter calyx than in the next species. FR, EP, NL; NSW, Vic.; native to Europe.

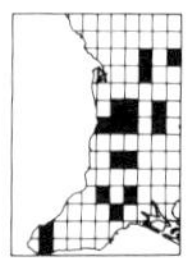

11. **Centaurium tenuiflorum*, **branched centaury**: Annual to 35 cm tall; lvs 0.5–3.0 cm long; fls (11a) 8–12 mm long when stretched out. LE, FR, EP, NL, MU, YP, KI, SE; all States except NT and Qld. **C. maritimum*, **sea centaury**, has yellow fls. **C. spicatum*, **spike centaury**, has fls along the branches; not confined to small terminal clusters.

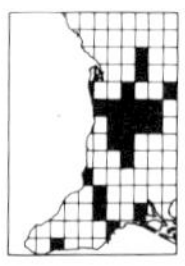

12. **Sebaea ovata*, **yellow sebaea**: Annual to 20 cm tall; lvs 3–12 mm long; fls (12a) 5–6 mm long when stretched out. FR, EP, NL, MU, YP, KI, SE; all States except NT.

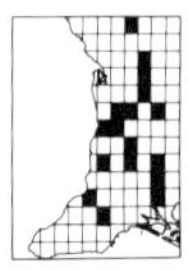

1
1a
2
2a
3
3a
3b
4
4a
4b
4c
5
5a
6
6a
6b
7
7a
7b
8
8a
9
9a
10
10a
11
11a
12
12a

Plate L

UMBELLIFERAE

1. *Centella cordifolia*, **centella**: Perennial herb with creeping stems; lf blades 1–3 cm diam., on stalks to 15 cm long; fls c. 1.5 mm long excl. ovary; fruit (1a) c. 2.5–3.0 mm long. FR, EP, MU, KI, SE; WA, Vic., Tas.

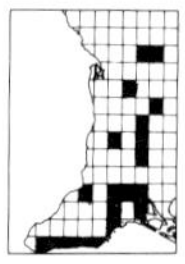

2. **Conium maculatum*, **hemlock**: Poisonous biennial herb 1.0–2.5 m tall; lvs to 50 cm long; fls (2a) 2–3 mm diam., fruit (2b) 2.5–3.5 mm long. EP, NL, MU, YP, KI, SE; all States except NT and Qld; New Zealand; native to Europe and Asia.

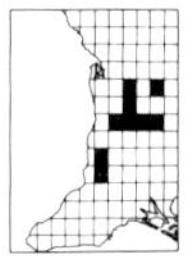

3. *Daucus glochidiatus*, **native carrot**: Annual herb usually 5–20 cm tall; lvs usually 1–8 cm long; fls (3a) c. 0.5 mm long excl. ovary; fruit (3b) 3–5 mm long; note barbed spines. All regions; all States; New Zealand. **D. carota*, **carrot**, is a larger plant with more infl branches.

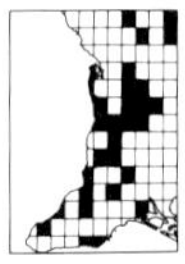

4. **Foeniculum vulgare*, **fennel**: Perennial herb 1.0–2.5 m tall; lvs to 50 cm long; fls c. 2 mm diam.; fruit (4a) 4.0–10.5 mm long. NL, MU, KI, SE; all States except NT; native to Europe.

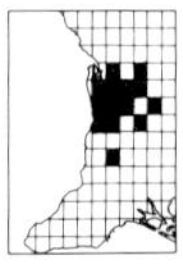

5. *Hydrocotyle callicarpa*, **tiny pennywort**: Annual herb 1–8 cm tall; lf blades 3–10 mm broad; fls minute; fruit (5a) usually to 1 mm long. FR, EP, NL, MU, YP, KI, SE; all States except NT and Qld.

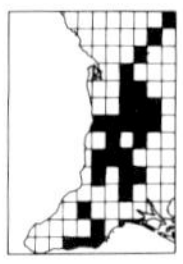

6. *Hydrocotyle foveolata*, **yellow pennywort**: Annual herb 1–10 cm tall; lf blades 3–9 mm broad; fls minute; fruit (6a) usually to 1 mm long. EP, NL, YP, KI, SE; NSW, Vic., Tas.

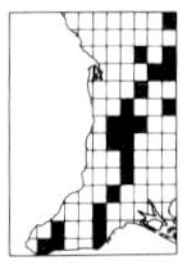

7. *Hydrocotyle laxiflora*, **stinking pennywort**: Perennial herb 6–16 cm tall; lf blades 1.0–4.5 cm broad; fls c. 1 mm long excl. ovary; fruit (7a) c. 2 mm long. FR, NL, KI, SE; all States except WA and NT.

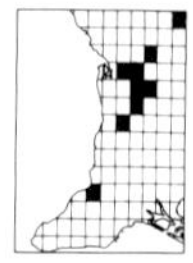

8. *Hydrocotyle verticillata*, **shield pennywort**: Perennial annual to 25 cm tall; lf blades 2–6 cm diam.; fls c. 1 mm long, excl. ovary; fruit (8a) c. 2 mm long. LE, FR, MU, SE; Qld, NSW, Vic., Tas. 7 other species.

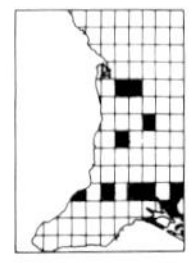

9. *Platysace heterophylla*, **slender platysace**: Perennial herb or shrub 10–30 cm tall; lvs 5–20 mm long; fls (9a) c. 1 mm long excl. ovary; fruit (9b) c. 1.5 mm long. EP, YP, KI, SE; Vic.

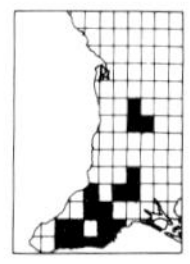

10. *Trachymene cyanopetala*, **purple trachymene**: Annual herb 5–25 cm tall; lf blades 15–20 mm diam., on stalks to 30 mm long; fls (10a) to 1 mm long excl. ovary; fruit (10b) 5–6 mm long. GT, EP, MU, YP, SE; all States except NT and Tas.

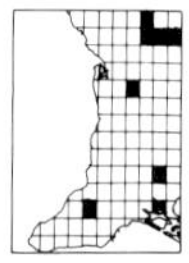

11. *Trachymene pilosa*, **dwarf trachymene**: Annual herb usually 3–15 cm tall; lf blades 15–25 mm diam., on stalks 5–50 mm long; fls (11a) c. 1 mm long excl. ovary; fruit c. 2 mm long. GT, FR, EP, NL, MU, YP, KI, SE; WA, Vic.

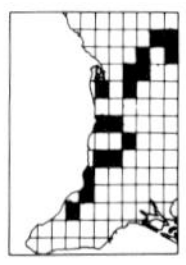

12. *Xanthosia pusilla*, **heath xanthosia**: Shrub 4–15 cm tall; lf segments 6–15 mm long, on stalks c. 6 mm long; fls (12a) c. 2 mm long excl. ovary; fruit (12b) 3–4 mm long. EP, KI, SE; WA, Vic., Tas. 2 other species.

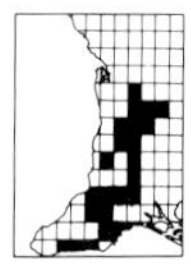

1
1a
2
2a
2b
3
3a
3b
4
4a
5
5a
6
6a
7
7a
8
8a
9
9a
9b
10
10a
10b
11
11a
12
12a
12b

Plate LI

EPACRIDACEAE

1. *Acrotriche affinis*, **prickly honeypots**: Shrub 20–100 cm tall; lvs 3.0–8.5 mm long; fls (1a) green or greenish-white, 4.0–4.5 mm long. NL, MU, YP, KI, SE; Vic.

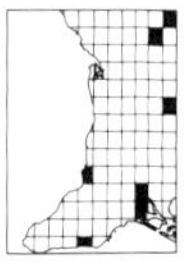

2. *Acrotriche depressa*, **native currant**: Shrub 40–80 cm tall; lvs 1.5–5.0 mm long; fls (2a) pale-green but largely concealed by bracts, 3.5–4.0 mm long; note infl of dense spikes. NL, MU, KI, SE; ?WA, Vic.

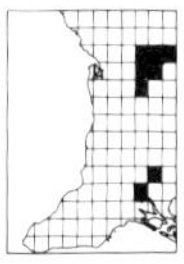

3. *Acrotriche fasciculiflora*, **pink ground-berry**: Shrub 50–150 cm tall; lvs (3a) 7–12 mm long; fls (3b) pale-pink, 7.0–10.5 mm long; note the dense arrangement of fls on the lower stems. KI.

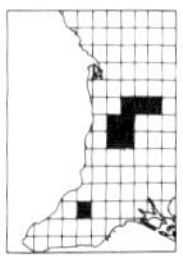

4. *Acrotriche patula*, **shiny-ground-berry**: Shrub 20–100 cm tall; lvs 6–16 mm long; fls (4a) pale-green, 3.2–5.2 mm long; fruit 4.0–6.3 mm diam. NU, EP, NL, MU, YP, KI; ?WA. *A. cordata*, **coast ground-berry**, has orange anthers and pale-green fruit.

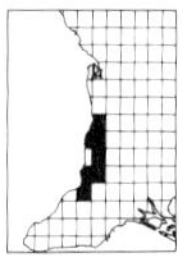

5. *Acrotriche serrulata*, **honeypots**: Mat-forming shrub 30–90 cm across; lvs 5–14 mm long; fls (5a) pale-green, 5.5–6.5 mm long. MU, KI, SE; NSW, Vic., Tas.

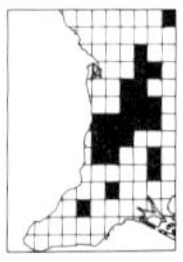

6. *Astroloma conostephioides*, **flame heath**: Shrub 30–100 cm tall; lvs 7–21 mm long; fls (6a) 12–20 mm long. EP, NL, MU, YP, SE; Vic.

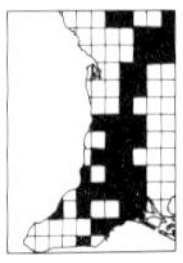

7. *Astroloma humifusum*, **native cranberry**: Mat-forming shrubs 30–40 cm across; lvs (7a) 6–18 mm long; fls (7b) 10.5–15.0 mm long. FR, EP, MU, YP, KI, SE; all States except NT and Qld.

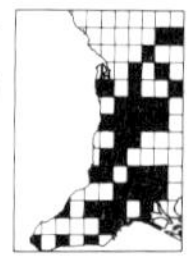

8. *Brachyloma ciliatum*, **fringed daphne heath**: Shrub 5–20 cm tall; lvs 4–11 mm long; fls (8a) white, 4–5 mm long; fruit (8b) 2.5–3.0 mm diam. ?KI, SE; Vic., Tas. *B. daphnoides*, **daphne heath**, has larger cream-coloured fls.

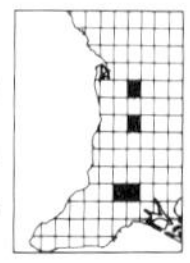

9. *Brachyloma ericoides*, **brush heath**: Shrub 20–90 cm tall; lvs 5–19 mm long; fls (9a) pink, 5–7 mm long. NL, MU, KI, SE; Vic.

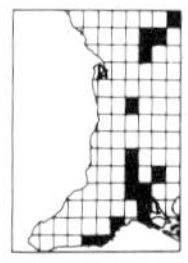

10. *Epacris impressa*, **native heath**: *Shrub 20–120 cm tall; lvs 4–13 mm long; fls (10a) red, pink or white, 8–21 mm long. FR, KI, SE; NSW, Vic., Tas.*

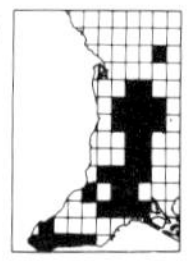

11. *Lissanthe strigosa*, **peach heath**: Shrub 30–100 cm tall; lvs 7–14 mm long; fls white (11a) or pinkish (11b), 3.5–6.0 mm long. FR, EP, NL, MU, KI, SE; all States except WA and NT.

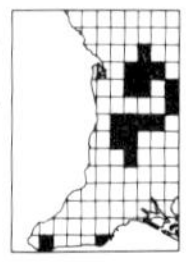

12. *Sprengelia incarnata*, **pink swamp-heath**: Shrub to 2 m tall; lvs 3–10 mm long; fls (12a) 4–6 mm long. KI, SE; NSW, Vic., Tas.

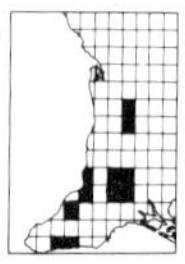

(*Continued*)

1a
1
2a
2
3b
3a
3
4a
4
5a
5
6a
6
7a
7b
7
8a
8b
8
9a
9
10a
10
11a
11b
11
12a
12

Plate LII

EPACRIDACEAE
(Continued)

1. *Leucopogon concurvus*, a **bearded-heath**: Shrub 30–60 cm tall; lvs (1a) concave above, 4–17 mm long; fls (1b) white, 3–4 mm long. KI.

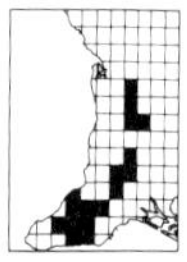

2. *Leucopogon cordifolius*, **heart-leaved bearded-heath**: Shrub 80–150 cm tall; lvs (2a) flat, 3–10 mm long; fls white, 5–7 mm long. GT, EP, NL, MU, YP, SE; WA, Vic.

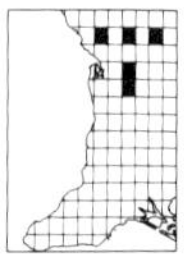

3. *Leucopogon parviflorus*, **coastal bearded-heath**: Shrub or tree 1.2–5 m tall; lvs (3a) flat or convex above, 11–29 mm long; fls white, 3.5–5.0 mm long; fruit (3b) 3.5–5.0 mm diam. EP, YP, KI, SE; all States except NT; New Zealand.

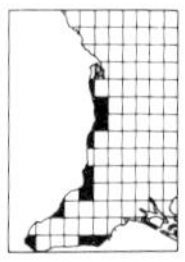

4. *Leucopogon rufus*, **ruddy bearded-heath**: Shrub 20–90 cm tall; lvs (4a) concave above, usually 4–18 mm long; fls pink, 4–7 mm long. EP, MU, KI, SE.

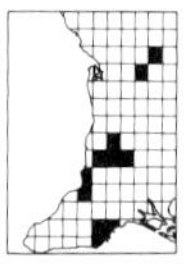

5. *Leucopogon virgatus*, **common bearded-heath**: Shrub 35–60 cm tall; lvs (5a) concave above, 3–21 mm long; fls white or pinkish, 3.5–4.7 mm long. NL, MU, YP, ?KI, SE; NSW, Vic., Tas. 5 other species.

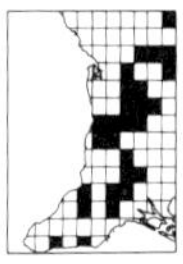

6. *Styphelia exarrhena*, **desert styphelia**: Shrub 30–60 cm tall; lvs thick, 1.5–5 mm long; fls (6a) white, 7.5–9 mm long. EP, MU, KI, SE; Vic.

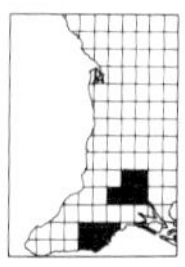

ARALIACEAE

7. *Hedera helix*, **ivy**: Creeping or climbing by adhering roots, with branches to 30 m long; lvs very variable in shape, to 15 cm diam.; fls 3–5 mm long. NSW, Vic., Tas.; native to Europe, Asia and N. Africa.

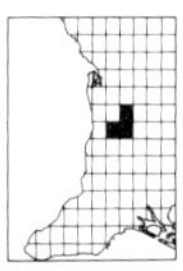

OLEACEAE

8. *Olea europaea*, **olive**: Tree usually 5–10 m tall; lvs 3–7 cm long; fls (8a) 4–5 mm long; fruit (8b) 1–3 cm long. FR, EP, NL, MU, YP, KI, SE; native to Europe, Asia and Africa.

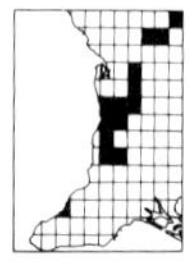

LOGANIACEAE

9. *Logania linifolia*, **flax-leaved logania**: Shrub to 1 m tall; stems minutely hairy; lvs 1–2 cm long; fls (9a) c. 2.5 mm long. MU, SE; Vic.

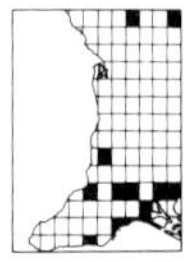

10. *Logania recurva*, **curved logania**: Shrub to 1 m tall; stems hairless; lvs usually 2–4 cm long; fls (10a) 2.3–3.5 mm long. NL, ?KI. 3 other species.

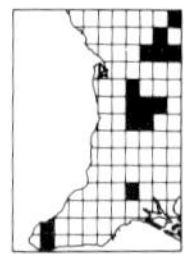

APOCYNACEAE

11. *Alyxia buxifolia*, **sea box**: Shrub to 2 m tall; lvs 2–4 cm long; fls (11a) white with an orange tube, c. 1.5 cm long; fruit (11b) c. 8 mm diam. Mainly on the coast. NU, GT, FR, EP, NL, MU, YP, KI, SE; all States except NT and Qld.

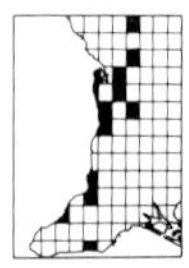

12. *Vinca major*, **periwinkle**: Weak perennial herb usually less than 1 m tall; lvs 2–7 cm long; fls (12a) with petals 2.0–2.5 cm long. NL, MU, YP, KI, SE; NSW, Vic., Tas.; native to the Mediterranean.

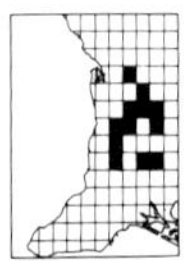

1
1a
1b
2
2a
3
3a
3b
4
4a
5
5a
6
6a
7
8
8a
8b
9
9a
10
10a
11
11a
11b
12
12a

MENYANTHACEAE

1. *Villarsia umbricola*, **lax marsh-flower**: Aquatic annual or perennial tufted herb rarely to 1 m tall, rooted in the mud; lvs erect or floating, the blades usually 1–8 cm long; fls (1a) 11–31 mm diam.; fruit (1b) 5–11 mm long. NL, KI, SE; Vic. *V. reniformis*, **running marsh-flower**, has leaves along stolons.

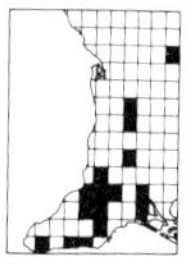

ASCLEPIADACEAE

2. **Asclepias fruticosa*, **narrow-leaved cotton-bush**: Shrub with milky sap, often c. 1 m tall; lvs c. 10 cm long; fls (2a) with petals 6–8 mm long; fruit (2b) to 6 cm long; seed (2c) c. 6 mm long excl. the long hairs. EP, MU, SE; Qld, NSW, Vic.; native to S. Africa. **A. physocarpus*, **balloon cotton-bush**, has a more rounded fruit.

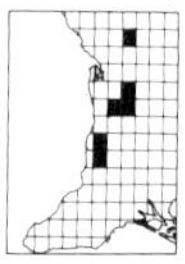

3. **Asclepias rotundifolia*, **broad-leaved cotton-bush**: Shrub with milky sap, c. 1 m tall; lvs to 6 cm long; fls (3a) with petals c. 8 mm long; fruit (3b) 7–8 cm long. NL, MU, YP, SE; Vic.; native to S. Africa.

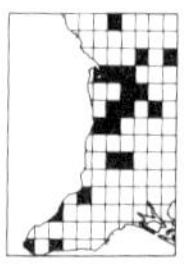

RUBIACEAE

4. *Asperula conferta*, **common woodruff**: Perennial herb to 20 cm tall; lvs 2–12 mm long; fls (4a male, 4b female) 1.5–3.5 mm long excl. ovary. FR, EA, EP, NL, MU, YP, SE; all States except WA and NT.

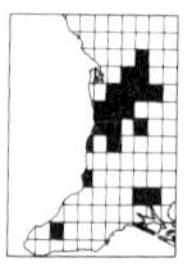

5. **Galium aparine*, **cleavers**: Weak-stemmed annual, often scrambling to more than 1 m; lvs 2–7 cm long; fls (5a) c. 1 mm long excl. ovary; fruit (5b) usually 4–5 mm long. SE; NSW, Vic., Tas.; native to Europe and Asia.

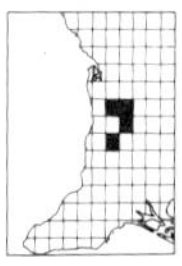

6. **Galium divaricatum*, **slender bedstraw**: Wiry annual to 30 cm tall; lvs 3–10 mm long; fls c. 0.5 mm long excl. ovary; fruit (6a) 0.5–1.0 mm long. NL, KI, SE; all States except NT and Qld; native to Europe and Asia.

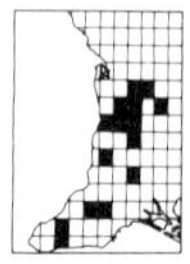

7. *Galium gaudichaudii*, **rough bedstraw**: Perennial to 30 cm tall; lvs 3–12 mm long, longer than the infl; fls (7a) c. 1 mm long excl. ovary; fruit (7b) 1.5–2.5 mm long. GT, FR, EA, EP, NL, MU, YP, SE; all States except NT and Qld.

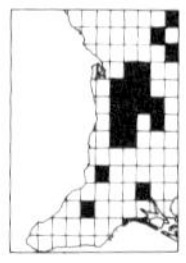

8. *Galium migrans*, a **bedstraw**: Perennial to 40 cm tall; lvs 4–15 mm long, shorter than the infl; fls c. 1 mm long excl. ovary; fruit (8a) to 2 mm long. NW, NU, FR, EA, EP, NL, MU, YP, KI, SE; all States except NT and Tas.

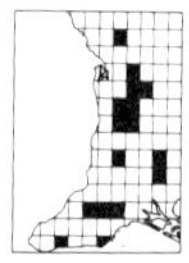

9. **Galium murale*, **small goosegrass**: Soft annual rarely more than 12 cm tall; lvs 3–10 mm long; fls (9a) c. 0.5 mm long excl. ovary; fruit (9b) 1.0–1.5 mm long, with characteristically hooked hairs. FR, EP, NL, YP, KI, SE; all States except NT and Qld; native to the Mediterranean. 3 other species.

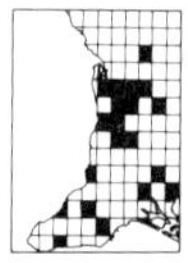

10. **Sherardia arvensis*, **field madder**: Annual to 30 cm tall; lvs 5–16 mm long; fls (10a) c. 3 mm long excl. ovary. EP, NL, MU, YP, KI, SE; NSW, Vic., Tas.; native to Europe and Asia.

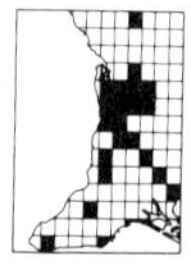

11. *Opercularia scabrida*, **stalked stinkweed**: Shrub to 25 cm tall; lvs 8–30 mm long; fls (11a) c. 2 mm long excl. ovary; fruit (11b) c. 4 mm long. EP, NL, MU, KI, SE; WA, NSW, Vic.

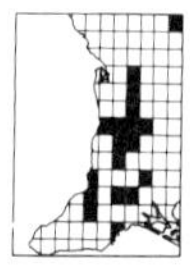

12. *Opercularia turpis*, **twiggy stinkweed**: Shrub to 30 cm tall; lvs 2–14 mm long; female fls (12a) c. 3 mm long excl. styles; male fls (12b) 3–4 mm long excl. anthers. FR, EP, NL, MU, YP, KI, SE; Vic.

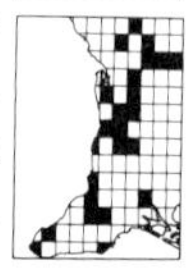

13. *Opercularia varia*, **variable stinkweed**: Shrub to 25 cm tall; lvs 3–18 mm long; fls (13a) c. 2 mm long excl. ovary. KI, SE; all States except WA and NT. *O. ovata*, **broad-leaved stinkweed**, has more fls (10–18) per infl.

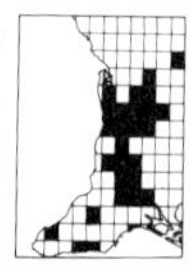

1a
1b
1
2a
2c
2
2b
3a
3b
3
4a ♂
4b ♀
4
5a
5b
5
6
6a
7a
7
7b
8a
8
9
9a
9b
10
10a
11a
11
11b
12 a ♀
12
12 b ♂
13
13a

CONVOLVULACEAE

1. *Calystegia sepium*, **greater bindweed**: Perennial vine to at least 4 m tall; lvs 4–10 cm long; fls white or pink, 4–6 cm long. MU, KI, SE; all States except NT and Qld.

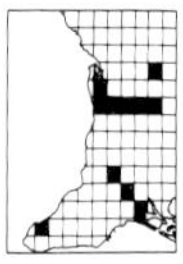

2. *Convolvulus erubescens*, **Australian bindweed**: Twining or trailing perennial to 2 m; lvs 15–55 cm long with spreading hairs; fls pink to reddish-purple, 7–11 mm long. All regions; all States. *C. remotus*, an **Australian bindweed**, has densely silvery-hairy sepals and with appressed hairs on the lvs.

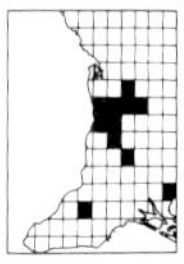

3. **Convolvulus arvensis*, **field bindweed**: Twining or trailing perennial to 2 m; lvs 15–50 cm long; fls white or pink, 15–25 mm long. FR, EP, MU, YP, SE; all States except NT; native to Europe and Asia.

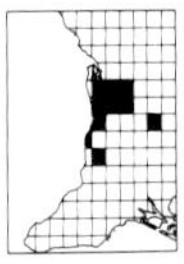

4. *Dichondra repens*, **Tom Thumb**: Perennial herb; lvs 5–80 mm long incl. stalk; fls (4a) 1.5–3.0 mm long. FR, EP, NL, MU, YP, KI, SE; all States except NT; southern Pacific.

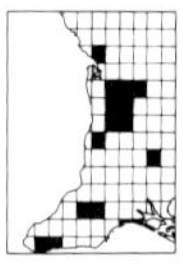

5. *Wilsonia humilis*, **silky wilsonia**: Prostrate shrub; lvs 1.5–4.0 mm long; fls (5a) 6–10 mm long. Grows in salty swamps. NU, EP, YP, KI, SE; WA, Vic., Tas.

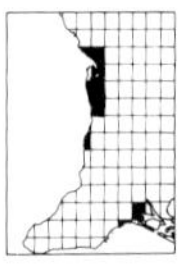

6. *Wilsonia rotundifolia*, **round-leaved wilsonia**: Prostrate shrub; lvs 1.5–4.0 mm long; fls (6a) 6.5–8 mm long. Grows in salty swamps. EP, NL, MU, YP, KI, SE; WA, NSW, Vic. *W. backhousei*, **narrow-leaved wilsonia**, has leaves at least 5 mm long.

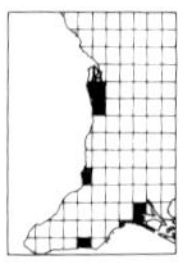

VERBENACEAE

7. **Phyla nodiflora*, **lippia**: Perennial creeping herb; lvs 1.0–7.2 cm long; fls (7a) purple to white, 2.0–2.5 mm across the petals. MU, YP; all States except Vic. and Tas.; warm parts of the World.

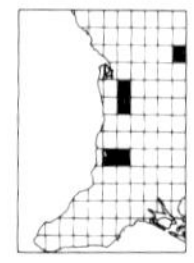

8. **Verbena bonariensis*, **purple-top**: Erect perennial to 2 m tall; lvs 4–22 cm long; calyx (8a) 2.5–3.5 mm long; fls blue, purple or pinkish, to 7 mm long. WA, Qld, NSW, Vic.; native to S. America. **V. rigida*, **veined verbena**, has bracts much longer than the calyx.

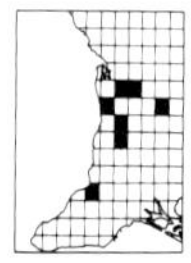

9. **Verbena officinalis*, **common verbena**: Erect perennial 30–90 cm tall; lvs 2–10 cm long; calyx (9a) 2.5–3.0 mm long; fls pale-pink to lilac, 5–7 mm long. NW, LE, FR, EA, MU, SE; all States except WA; native to the Mediterranean.

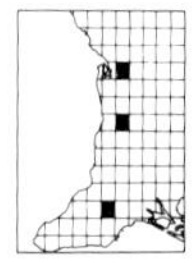

10. **Verbena supina*, **trailing verbena**: Prostrate herb with branches 15–50 cm long; lvs 2–4 cm long; calyx (10a) c. 3 mm long; fls (10b) lilac, to 6 mm long. LE, NU, GT, FR, EA, EP, NL, MU, YP, SE; NSW, Vic.; native to the Mediterranean. **V. tenuisecta*, **Maynes pest**, has a narrow calyx 6–9 mm long.

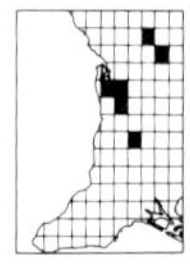

AVICENNIACEAE

11. *Avicennia marina*, **mangrove**: Tree or shrub 2–10 m tall; lvs leathery, 3–8 cm long; fls (11a) white to orange-yellow, 4–6 mm long; coastal, growing in the tidal area mainly. EP, NL, YP; all mainland States; western Pacific.

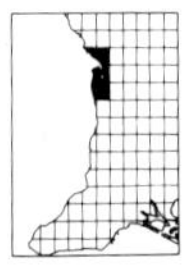

CALLITRICHACEAE

12. **Callitriche stagnalis*, **common starwort**: Mainly aquatic herb; lvs 2.5–25.0 mm long, 3-nerved; male and female fls lacking petals. NL, SE; all States; almost cosmopolitan. *C. umbonata*, a **water starwort**, has very narrow 1-nerved lvs.

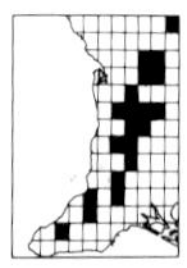

1
2
3
5a
6a
4a
4
5
6
8a
9a
7a
7
8
9
10b
11a
10a
10
11
12
♀
♂

Plate LV

BORAGINACEAE

1. **Buglossoides arvensis*, **sheepweed**: Annual herb to 30 cm tall; lvs 1-7 cm long; fls (1a) white to cream, 6-7 mm long; part fruit (1b) 3-4 mm long. LE, NU, FR, EA, EP, NL, MU, YP, KI, SE; WA; native to N. Africa, Europe and Asia.

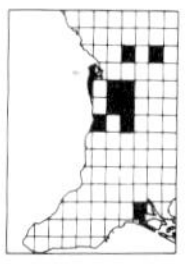

2. *Cynoglossum australe*, **Australian hounds tongue**: Annual herb to 60 cm tall; lvs 2-25 cm long; fls (2a) blue, 4-6 mm long; part fruit (2b) c. 5 mm long, saucer-shaped, winged. NW, LE, FR, EP, YP, KI, SE; all States.

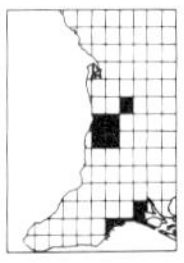

3. *Cynoglossum suaveolens*, **sweet hounds tongue**: Perennial herb to 40 cm tall; lvs 2-15 cm long; fls (3a) blue or white, 4.0-4.5 mm long; part fruit (3b) 2.0-3.5 mm long, almost spherical; note small leaf at the base of each fl. FR, EP, NL, SE; all States except WA and NT.

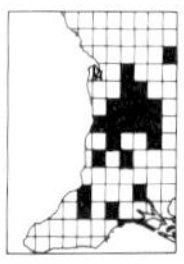

4. **Echium plantagineum*, **Salvation Jane**: Annual or biennial herb to 80 cm tall; lvs 4-20 cm long; fls (4a and 4b) 18-30 mm long. NW, LE, NU, GT, FR, EA, EP, NL, MU, YP, SE; all States; native to N. Africa and Europe. **E. italicum*, **Italian bugloss**, has cream or pale-yellow fls.

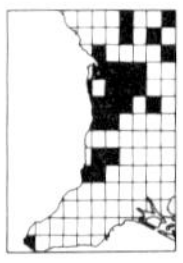

5. *Halgania andromedifolia*, **lavender halgania**: Shrub to 1.3 m tall; lvs 1.0-2.5 cm long; fls (5a) 5-8 mm long. GT, EP, NL, MU, YP; WA, NSW, Vic. *H. cyanea*, **rough halgania**, has double-ended hairs on the sometimes toothed lvs.

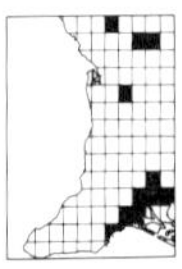

6. **Heliotropium amplexicaule*, **blue heliotrope**: Perennial herb to 30 cm tall, with branches on the ground; lvs 2.5-8.0 cm long; fls (6a) lilac to purple, c. 6.0-7.5 mm across the petals. GT, FR, MU, SE; Qld, NSW, Vic.; native to S. America.

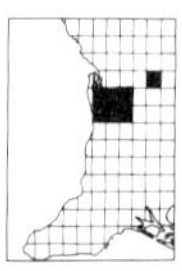

7. **Heliotropium europaeum*, **potato weed**: Erect annual herb 10-50 cm tall; lvs 1.5-9.0 cm long; fls (7a) white, c. 3.5-6.0 mm across the petals; fruit (7b) 2-3 mm long. LE, NU, FR, EA, EP, NL, MU, YP, SE; WA, NSW, Vic.; native to Europe. 2 other species with white or pinkish fls.

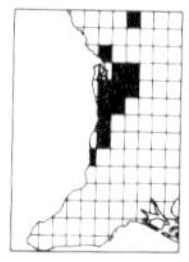

8. *Myosotis australis*, **austral forget-me-not**: Annual herb to 25 cm tall; lvs 0.8-4.0 cm long; fls (8a) blue or white, 2-4 mm long; fruit (8b) c. 2 mm long. GT, EP, NL, YP, KI, SE; all States except NT and Qld; New Zealand, New Guinea. 2 other species.

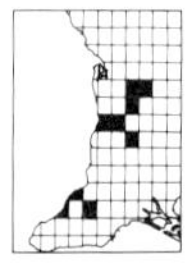

9. **Neatostema apulum*, **hairy sheepweed**: Annual herb to 25 cm tall; lvs 1.0-3.5 cm long; fls yellow, 4-5 mm long. EP, MU, NL, YP, SE; Vic.; native from the Canary Islands to Iraq.

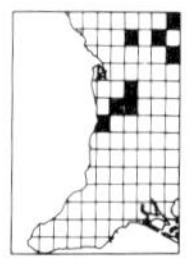

LABIATAE

10. *Lycopus australis*, **Australian gypsywort**: Perennial herb to 1.5 m tall; lvs 4-15 cm long; fls (10a) white, 4.5-5.5 mm long. Grows in wet places. MU, SE; all States except WA and NT.

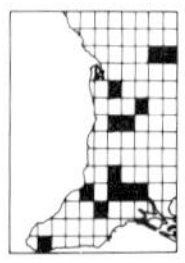

11. **Marrubium vulgare*, **horehound**: Shrub to 60 cm tall; lvs 1-7 cm long; fls (11a and 11b) white, 6-12 mm long. NU, GT, FR, EA, EP, NL, MU, YP, SE; all States except NT; New Zealand; native to Europe.

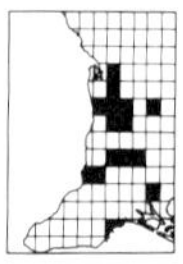

12. *Teucrium racemosum*, **grey germander**: Perennial herb to 60 cm tall; lvs 1-5 cm long; fls (12a) white, 9-10 mm long when stretched out. NW, LE, GT, FR, EA, EP, NL, MU, YP, SE; all mainland States. *T. sessiliflorum*, **camel bush**, has lobed lvs.

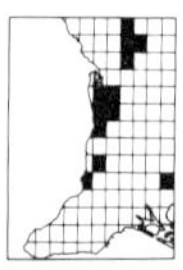

(*Continued*)

1a
1b
1
2b
2a
2
3b
3a
3
4a
4b
4
5a
5
6a
6
7a
7b
7
8a
8b
8
9a
9
10a
10
11b
11a
11
12a
12

Plate LVI

LABIATAE
(*Continued*)

1. *****Lavandula stoechas*, **topped lavender**: Shrub rarely to 80 cm tall; lvs 1–3 cm long; fls (1a) 7–8 mm long. NL, MU; WA, Qld, NSW, Vic.; native to Europe. **L. dentata*, **French lavender**, has shallowly lobed lvs.

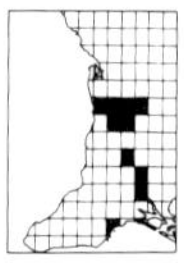

2. **Melissa officinalis*, **balm**: Perennial herb to 1 m tall; lvs 3–12 cm long; fls (2a) pale-yellow to pinkish-blue, 8–12 mm long. NSW, Vic., Tas.; probably native to the Mediterranean.

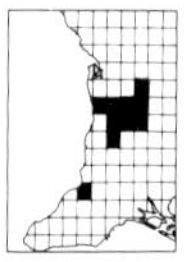

3. *Mentha diemenica*, **slender mint**: Perennial herb with more or less prostrate branches to 25 cm long; stem section (3a); lvs 1.0–2.3 cm long; fls (3b) 5–7 mm long. NL, SE; all States except WA and NT.

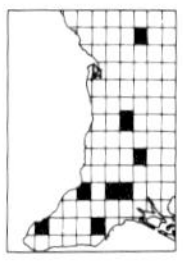

4. **Mentha piperita*, **lemon mint**: Perennial herb to 80 cm tall; lvs 1.5–9.5 cm long; fls lilac to pink, 4–7 mm long. NL, MU, SE; NSW, Vic.; a hybrid between 2 Northern Hemisphere species. Included in this species is the peppermint, which has narrower lvs.

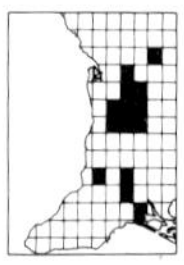

5. **Mentha spicata*, **spearmint**: Perennial herb to 1 m tall; lvs 2–9 cm long; fls lilac, pink or white, 3–4 mm long. MU, YP, SE; WA, NSW, Vic.; native to the Mediterranean. 4 other species.

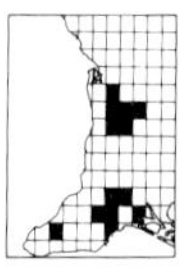

6. *Prostanthera aspalathoides*, **scarlet mintbush**: Shrub 0.3–1.0 m tall; lvs usually 1.5–6.0 cm long; fls red to orange or yellow, 13–19 mm long. EP, MU, YP, KI, SE; Qld, NSW, Vic.

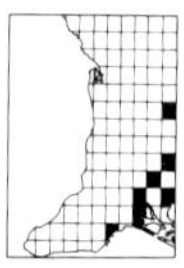

7. *Prostanthera behriana*, **downy mintbush**: Shrub 1.0–2.5 m tall; lvs usually 14–26 mm long; fls (7a) white to pale-blue or pale-purple, 13–20 mm long. FR, EP, NL, MU, KI, SE.

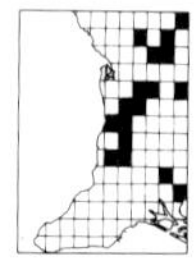

8. *Prostanthera chlorantha*, **green mintbush**: Shrub rarely to 1 m tall; lvs 1.0–3.5 mm long; fls (8a) mauve, blue-green or green-yellow, 18–26 mm long. EP, MU, YP, KI. 3 other species.

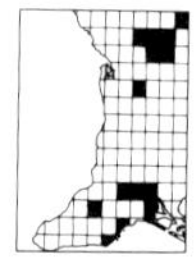

9. *Prunella vulgaris*, **self-heal**: Perennial herb with more or less prostrate branches to 40 cm long; lvs 2–8 cm long; fls (9a) 8–12 mm long. ?MU; all States except NT.

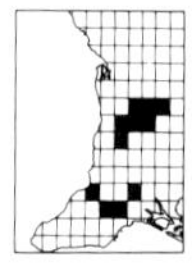

10. **Salvia verbenaca*, **wild sage**: Perennial herb to 70 cm tall; lvs (10a) 4–18 cm long; fls (10b) bluish-purple to lilac, 8–13 mm long; fruiting calyx (10c) to 9 mm long. FR, EP, NL, MU, YP, KI, SE; all States; native to Europe and N. Africa. 2 other species.

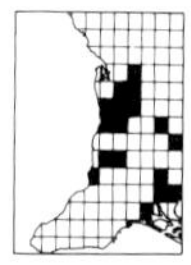

11. **Stachys arvensis*, **stagger weed**: Annual herb to 35 cm tall; lvs 1–7 cm long; fls (11a) white or pale-pink, 6–8 mm long. MU, SE; all States except NT; New Zealand; native to Europe and N. Africa. *S. byzantina*, **woolly stachys**, is a woolly perennial.

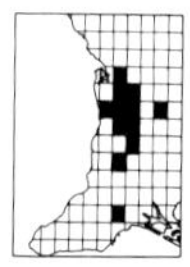

12. *Westringia rigida*, **stiff westringia**: Shrub rarely to 1 m tall; lvs 1.9–5.2 mm long; fls (12a) 6–7 mm long. NW, NU, GT, FR, EA, EP, NL, MU, YP, SE; WA, NSW, Vic. *W. eremicola*, **slender westringia**, has lvs with stalks c. 0.5 mm long and usually more deeply mauve fls.

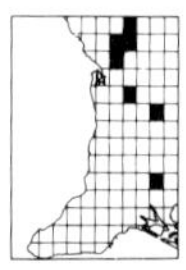

1a
1
2a
2
3a
3b
3
4
5
6
7a
7
8a
8
9a
9
10c
10b
10a
10
11a
11
12a
12

PLATE LVII

SOLANACEAE

1. *Datura stramonium*, **common thornapple**: Annual herb 0.2–1.2 m tall; lvs 7–36 cm long; fls white or pale-lavender, usually 6.5–8.5 cm long; fruit (1a) erect, 2.0–4.5 cm long; seed (1b) 2.5–4.5 mm long. FR, EA, EP, NL, MU, YP, KI, SE; all States except NT; probably native to N. America. **D. ferox*, **fierce thornapple**, has similar lvs but fewer thicker spines on the fruit.

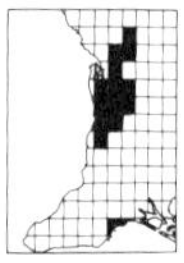

2. *Datura wrightii*, **hairy thornapple**: Usually perennial herb 0.7–1.2 m tall; lvs 5–16 cm long; fls white, sometimes tinged lavender, 14–20 cm long; fruit (2a) pendulous, usually 2.5–3.5 cm long; seed (2b) 5.0–5.8 mm long. FR, MU; all mainland States except probably NT; native to N. America. **D. metel*, a **thornapple**, has similar lvs but blunt tubercles on the fruit and occurs as a rare garden escape.

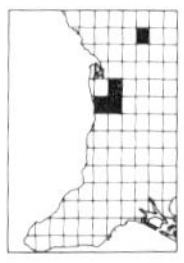

3. *Lycium ferocissimum*, **African boxthorn**: Shrub usually 1.5–2.5 m tall; lvs 1–3 cm long; fls pale-lilac or whitish, 10–12 mm long; fruit (3a) 5–12 mm diam. FR, EA, EP, NL, MU, YP, KI, SE; all States; native to S. Africa. 2 other species.

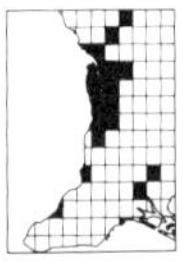

4. *Nicotiana glauca*, **tree tobacco**: Shrub or tree to 6 m tall; lvs usually to 20 cm long; fls 25–50 mm long; fruit (4a) 7–13 mm long. LE, NU, GT, FR, EA, EP, NL, MU, YP, KI, SE; all mainland States; native to S. America.

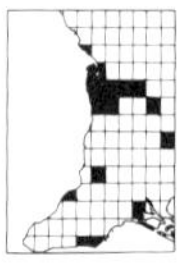

5. *Nicotiana maritima*, **coast tobacco**: Herb usually to 0.7 m tall; lvs mostly basal, usually to 22 cm long; fls (5a) usually 2–3 cm long. FR, EP, NL, MU, YP, KI, SE; ?Vic. *N. goodspeedii*, **small-flowered tobacco**, lacks hairs.

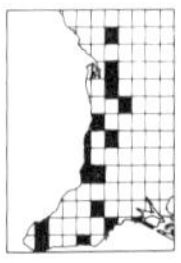

6. *Solanum elaeagnifolium*, **silver-leaved nightshade**: Perennial herb to 1 m tall; lvs to c. 10 cm long; fls (6a) 2–3 cm diam.; fruit (6b) becoming yellowish or orangish, 8–14 mm diam. EP, NL, MU, YP, SE; all mainland States; native to N. America.

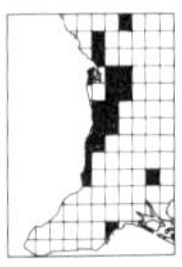

7. *Solanum linnaeanum*, **apple of Sodom**: Shrub to 1 m tall; lvs to 8 cm long; fls c. 2 cm diam.; fruit 2–3 cm diam. EP, MU, YP, KI, SE; all mainland States except NT; native to Europe and Africa.

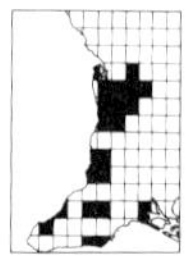

8. *Solanum laciniatum*, **kangaroo apple**: Shrub 1–3 m tall; lvs 5–30 cm long, very variable in shape; fls (8a) 3–5 cm diam.; fruit (8b) yellow or orange-yellow, 1.5–2.0 cm diam. EP, NL, KI, SE; WA, Vic., Tas.

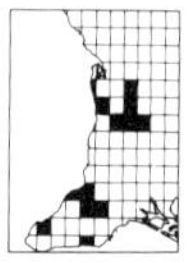

9. *Solanum mauritianum*, **wild tobacco tree**: Shrub or tree 3–4 m tall; lvs usually 10–30 cm long; fls 1.5–2.5 cm diam.; fruit (9a) 1.0–1.5 cm diam. Qld, NSW; native to S. America.

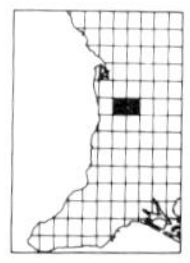

10. *Solanum nigrum*, **black-berry nightshade**: Herb or shrub to 1 m tall; lvs 4–7 cm long; fls c. 1 cm diam.; fruit (10a) c. 6–8 mm diam. LE, NU, GT, FR, EA, EP, NL, MU, YP, KI, SE; all States; native to Europe. 6 other species.

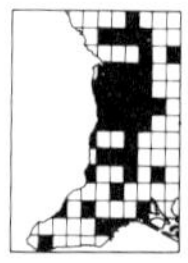

SCROPHULARIACEAE

11. *Kickxia commutata*, a **toadflax**: Prostrate perennial herb; lvs to 2 cm long; fls (11a) blue, violet or purple, 5–8 mm long; note the hard-walled fruit. NSW, ?Vic.; native to the Mediterranean.

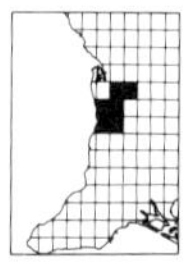

12. *Kickxia elatine*, **pointed toadflax**: Prostrate perennial herb; lvs 1–3 cm long; fls (12a) purplish, yellow and white, 2.5–7.5 mm long; note the fragile fruit. LE, FR, EP, NL, MU, KI, SE; NSW, Vic.; native to Europe. **K. spuria*, **round-leaved toadflax**, lacks the lobes on either side of the lvs.

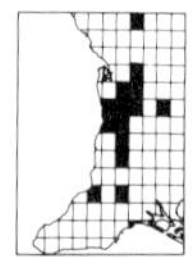

(*Continued*)

1
1a
1b
2
2a
2b
3
3a
4
4a
5
5a
6
6a
6b
7
8
8a
8b
9
9a
10
10a
11
11a
12
12a

LENTIBULARIACEAE

1. *Utricularia dichotoma*, **purple bladderwort**: Aquatic herb 10–50 cm tall, with 1–5 mm long bladder-like traps on the creeping stems; lvs 0.3–10.0 cm long; fls 1–2 cm long. In wet places. FR, KI, SE; all States except NT. 2 other species.

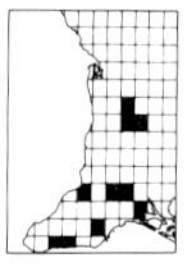

SCROPHULARIACEAE

(*Continued*)

2. *Gratiola peruviana*, **austral brooklime**: Weak or erect perennial herb usually 9–50 cm tall; lvs usually 1.5–4.5 cm long; fls (2a) pale-pink, usually 10–12 mm long. MU, KI, SE; all States except NT and Qld. *G. pumilo*, a brooklime, has white fls.

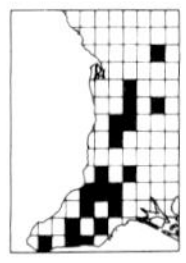

3. *Mimulus repens*, **creeping monkey-flower**: Annual or perennial virtually hairless herb to 20 cm tall; lvs usually 1–4 mm long; fls (3a and 3b) purple or pink, 2.5–5.0 mm long. Sometimes in wet places. LE, GT, EP, NL, MU, KI, SE; all States except NT; New Zealand.

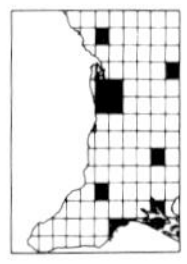

4. *Parentucellia viscosa*, **yellow bartsia**: Annual herb 10–70 cm tall; lvs 10–42 mm long; fls (4a) 15–24 mm long. SE; all States except NT and Qld; native to the Mediterranean.

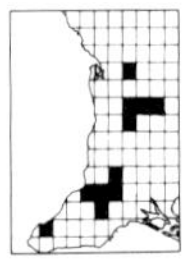

5. *Mimulus moschatus*, **musk monkey-flower**: Perennial herb with the main branches to 50 cm long; lvs 22–70 mm long; fls yellow, c. 2 cm long. In wet places. NSW, Vic., Tas.; New Zealand; native to N. America.

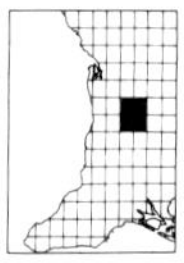

6. *Misopates orontium*, **lesser snapdragon**: Annual herb 15–60 cm tall; lvs 1.5–6.5 cm long; fls (6a) 10–14 mm long; fruit (6b) 6–8 mm long. FR, EP, NL, KI, SE; WA, Qld, NSW; Vic.; probably native to the Mediterranean.

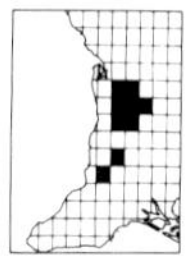

7. *Parentucellia latifolia*, **red bartsia**: Annual herb 5–30 cm tall; lvs 3–15 mm long; fls (7a) 8–10 mm long. EP, NL, MU, YP, KI, SE; WA, NSW, Vic.; native to the Mediterranean.

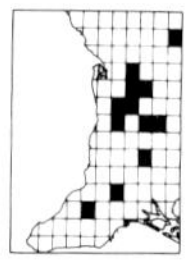

8. *Polycarena heterophylla:* Annual herb 3–20 cm tall; lvs 5–35 mm long; fls (8a) c. 3 mm long. EP, MU; WA; native to S. Africa. *P. leipoldtii* is glandular-hairy and has larger fls in a more open infl.

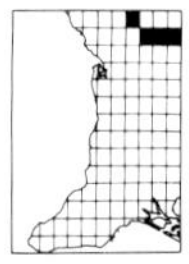

9. *Verbascum virgatum*, **twiggy mullein**: Biennial herb to 3 m tall; lvs 8–30 cm long; fls (9a) 3–4 cm diam. FR, EP, SE; all States except NT; native to Europe. 2 other introduced species.

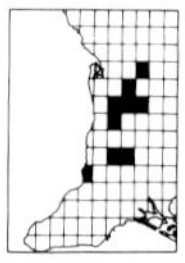

10. *Veronica hillebrandii*, **coast speedwell**: Perennial herb 4–32 cm tall; lvs 5–30 mm long; fls 4.8–10.0 mm long. EP, MU, YP, KI.

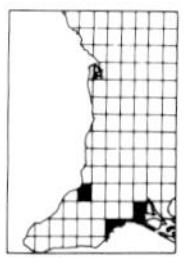

11. *Veronica persica*, **creeping speedwell**: Weak annual 6–50 cm tall; lvs 5–30 mm long; fls (11a) 3–5 mm long. SE; all States except NT and Qld; native to Asia. 4 other species.

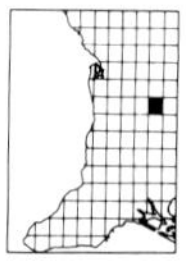

12. *Zaluzianskya divaricata*, **spreading night-phlox**: Annual 3–35 cm tall; lvs 1–5 cm long; fls (12a) 17–23 mm long. EP, ?NL, MU, YP, KI, SE; WA, Vic.; native to S. Africa.

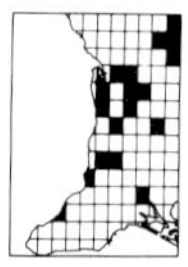

1
2
2a
3
3a
3b
4
4a
5
6
6a
6b
7
7a
8
8a
9
9a
10
11
11a
12
12a

PLATE LIX

MYOPORACEAE

1. *Eremophila longifolia*, **weeping emubush**: Shrub or tree 1-8 m tall; lvs 3-20 cm long; fls (1a); petals 2-3 cm long when stretched out. NW, LE, NU, GT, FR, EA, EP, NL, MU, YP; all mainland States. *E. subfloccosa*, an **emubush**, also has hairs on the fls which are yellow to green.

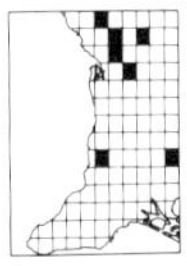

2. *Eremophila deserti*, **turkey-bush**: Shrub 1-4 m tall; lvs 2.0-5.5 cm long; petals 4-7 mm long. NW, LE, NU, GT, FR, EA, EP, NL, MU, YP; all States except NT and Tas.

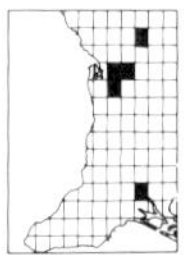

3. *Eremophila gibbifolia*, **coccid emubush**: Shrub 0.3-0.9 m tall; lvs 1.5-5.0 mm long; petals 7.0-10.5 mm long. EP, MU; Vic. *E. behriana*, **rough emubush**, also has petals hairless outside but the lvs are up to 2.6 cm long.

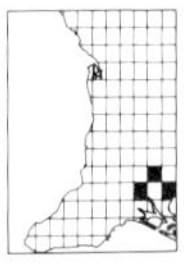

4. *Myoporum montanum*, **native myrtle**: Shrub or tree 1-4 m tall; lvs (4a) 3-14 cm long; fls (4b) 5-8 mm long when stretched out; fruit 6-8 mm diam. NW, LE, NU, GT, FR, EA, EP, NL, MU, YP; all mainland States.

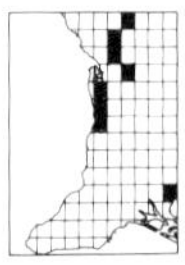

5. *Myoporum insulare*, **native juniper**: Shrub 1-3 m tall; lvs (5a) 3-7 cm long; fls (5b) 3-6 mm long when stretched out; fruit 4-8 mm diam. NU, EP, YP, KI, SE; WA, NSW, Vic.

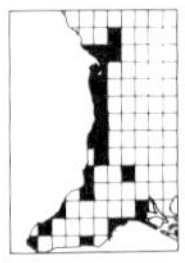

6. *Myoporum parvifolium*, **creeping boobialla**: Mat-forming shrub usually less than 10 cm tall; lvs 0.5-4.0 cm long; fls (6a) 3-6 mm long when stretched out; fruit 5-7 mm diam. EP, MU, YP, KI, SE; Vic.

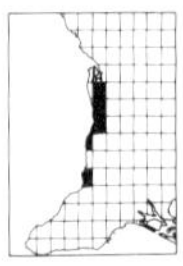

7. *Myoporum platycarpum*, **false sandalwood**: Shrub or tree 1-10 m tall; lvs (7a) 3-9 cm long; fls (7b) 3.5-7.0 mm long when stretched out; fruit flattened, 4-5 mm long excl. the style. NW, LE, NU, GT, FR, EA, EP, NL, MU, YP, SE; WA, NSW, Vic.

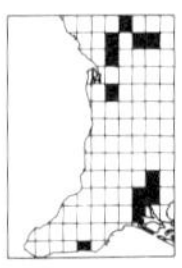

8. *Myoporum viscosum*, **sticky boobialla**: Shrub to 2 m tall; lvs 2-11 cm long; fls (8a) 6-7 mm long when stretched out; fruit c. 5 mm diam. FR, EP, NL, MU, YP, KI, SE; Vic. *M. brevipes*, a **myoporum**, has (like *M. viscosum* and *M. platycarpum*) strongly tuberculate branches but usually smooth margins to the lvs.

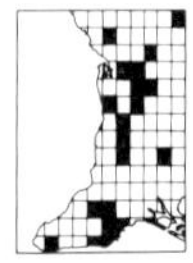

VALERIANACEAE

9. **Centranthus macrosiphon:* Annual herb to 40 cm tall; lvs 1.5-4.0 cm long; fls (9a) 6-8 mm long, with a basal spur (not shown) c. 1 mm long. SE; WA; native to Spain.

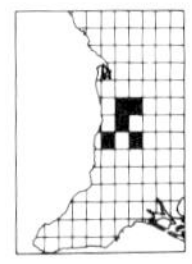

10. **Centranthus ruber*, **red valerian**: Perennial herb to 70 cm long; lvs (10a) 3.5-12.0 cm long; fls (10b) 7-11 mm long, with a basal spur 4-8 mm long. YP; all States except NT and Qld; native to Europe.

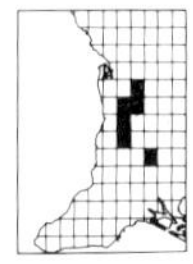

11. **Valerianella muricata*, **narrow-fruit cornsalad:** Annual herb to 20 cm tall; lvs 1.5-8.0 cm long; fls (11a) c. 2 mm long; fruit (11b) c. 2 mm long. SE; native to southern Europe and central Asia. 2 other species.

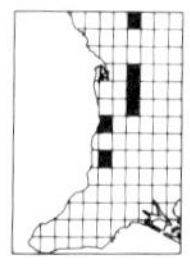

DIPSACACEAE

12. **Scabiosa atropurpurea*, **scabious**: Perennial herb to 70 cm tall; lvs 3-15 cm long; fls (12a) in white to purple heads (12b) 1.5-3.0 cm diam. FR, EP, NL, MU, YP, KI, SE; all States except NT and Qld.

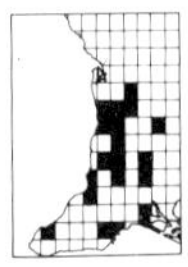

1
1a
2
3
4
4a
5
5a
5b
6
6a
7
7a
7b
8
8a
9
9a
10
10a
10b
11
11a
11b
12
12a
12b

PLANTAGINACEAE

1. **Plantago coronopus*, **bucks-horn plantain**: Annual or perennial 2–30 cm tall; lvs 2–35 cm long; flower-spike (1a) usually 1–5 cm long; note the tap root and dense flower-spike. LE, EA, EP, NL, MU, YP, KI, SE; all States except NT; native to N. Africa, Europe and Asia.

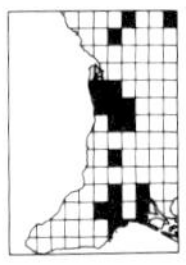

2. *Plantago gaudichaudii*, **narrow-leaved plantain**: Perennial 5–40 cm tall; lvs (2a) 8–28 cm long; flower-spike 2–10 cm long; note the tap root and leaves with rather few hairs. GT, FR, EP, NL, YP, SE; all States except WA and NT.

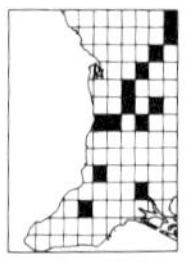

3. **Plantago lanceolata*, **ribgrass**: Annual or perennial 10–50 cm tall; lvs (3a) 4–22 cm long; flower-spike (3b) usually 8–25 cm long; note usual absence of tap root. FR, NL, MU, YP, KI, SE; all States except NT; New Zealand; native to N. Africa, Europe and Asia.

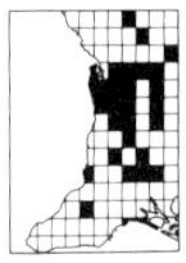

4. **Plantago major*, **greater plantain**: Perennial 10–50 cm tall; lvs 4–47 cm long; flower-spike (4a) usually 8–25 cm long; note absence of tap root. MU, SE; all States except NT; New Zealand; native to the Northern Hemisphere.

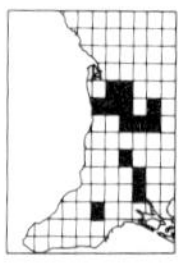

5. *Plantago varia*, **variable plantain**: Perennial 6–40 cm tall; lvs 6–25 cm long; flower-spike (5a) 1.4–14.0 cm long; note the tap root. EP, MU, SE; NSW, Vic., Tas. 6 other species.

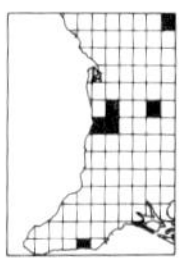

CAMPANULACEAE

6. *Lobelia alata*, **angled lobelia**: Weak perennial herb to 50 cm tall with angled stems (6a); lvs 15–70 mm long; fls (6b and 6c) 5–8 mm long when stretched out excl. ovary, on hairy flower-stalks. FR, EP, NL, MU, KI, SE; all States except NT; New Zealand, Africa, S. America.

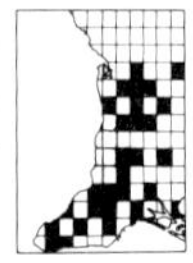

7. *Lobelia gibbosa*, **tall lobelia**: Erect annual herb to 30 cm or more tall; lvs 20–70 mm long; fls (7a) 10–15 mm long when stretched out excl. ovary, on hairless flower-stalks. FR, EP, NL, MU, YP, KI, SE; all States except NT.

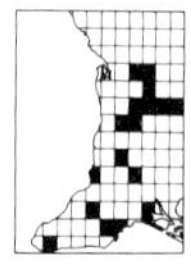

8. *Lobelia rhombifolia*, **tufted lobelia**: Weak annual herb to 20 cm or more tall; lvs 8–15 mm long; fls (8a) 11–14 mm long when stretched out excl. ovary, on hairless flower-stalks. YP, KI, SE; WA, Vic., Tas.

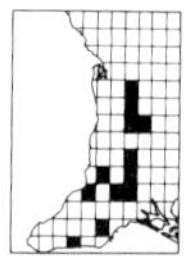

9. *Wahlenbergia communis*, **tufted bluebell**: Perennial herb 5–75 cm tall; lvs alternate, 4–80 mm long; fls (9a and 9b) blue or white, 10–22 mm long when stretched out excl. ovary. NW, LE, GT, FR, EA, EP, NL, MU, YP, SE; all mainland States; New Guinea. *W. gracilis*, **sprawling bluebell**, also has alternate lvs but smaller fls.

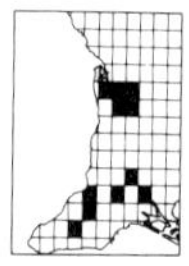

10. *Wahlenbergia gracilenta*, **annual bluebell**: Annual herb 1.5–35.0 cm tall; lvs opposite below but alternate above, 1–40 mm long; fls (10a) blue, white or pink, 1–9 mm long when stretched out excl. ovary; fruit (10b). LE, GT, FR, EA, EP, NL, MU, YP, KI, SE; all States except NT and Qld.

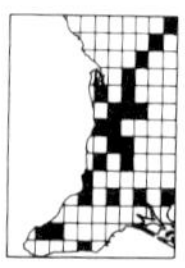

11. *Wahlenbergia luteola*, a **bluebell**: Perennial herb 6–80 cm tall; lvs opposite below but alternate above, 4–60 mm long; fls (11a and 11b) blue, pink or white, 8–19 mm long excl. ovary; fruit (11c). LE, FR, EP, NL, MU, YP, SE; NSW, Vic.

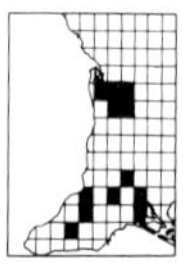

12. *Wahlenbergia stricta*, **tall bluebell**: Perennial herb 10–90 cm tall; lvs opposite below but alternate above, 5–70 mm long; fls (12a) blue or white, 10–30 mm long excl. ovary. GT, FR, EP, NL, MU, YP, SE; all States except NT; New Zealand. 3 other species.

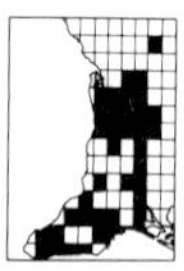

1a
1
2
2a
3
3a
3b
4
5
5a
6
6a
6b
6c
7
7a
8
8a
9
9a
9b
10
10a
10b
11
11a
11b
11c
12
12a

Plate LXI

GOODENIACEAE

1. *Brunonia australis*, **blue pincushion**: Perennial herb 6–32 cm tall; lvs 4–10 cm long; heads 1.4–2.4 cm diam.; fls (1a) with petals 3–4 mm long. NW, LE, GT, SE; all States.

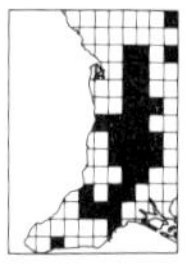

2. *Dampiera dysantha*, a **dampiera**: Shrub to 70 cm tall; lvs 5–24 mm long, usually scattered along the stems; fls (2a) blue, lilac or white, 11–18 mm long excl. ovary; bracts at base of fls 5–7 mm long. FR, EP, NL, MU, SE; Vic.

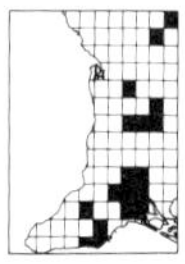

3. *Dampiera lanceolata*, **grooved dampiera**: Shrub to 1 m tall; lvs 5–47 mm long, usually scattered along the stems; fls (3a) blue, 11–14 mm long excl. ovary; bracts at base of fls 2–4 mm long. NW, NU, GT, EP, NL, MU, KI, SE; Qld, NSW, Vic.

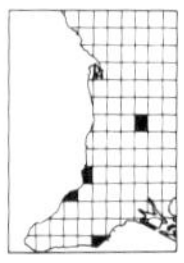

4. *Dampiera rosmarinifolia*, **wild rosemary**: Shrub to 60 cm tall; lvs 8–28 mm long, often in clusters along the stems; fls (4a and 4b) blue-violet to pink, 12–14 mm long excl. ovary. NU, EP, NL, MU, YP, SE; Vic. 2 other species.

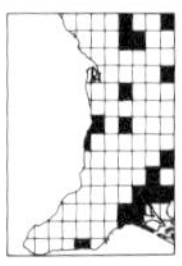

5. *Goodenia albiflora*, **white goodenia**: Herb to 50 cm tall; lvs along the stem, 3–7 cm long; fls (5a) 15–20 mm long excl. ovary. FR, EA, NL, MU.

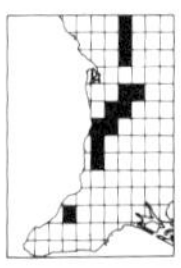

6. *Goodenia amplexans*, **clasping goodenia**: Shrub to 1 m tall; lvs along the stem, 3–9 cm long; fls (6a) 12–15 mm long excl. ovary; note the shape of the lf bases. FR, EP, NL, KI; Vic.

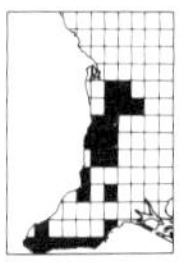

7. *Goodenia blackiana*, a **goodenia**: Perennial herb with stems to 20 cm long; lvs in basal rosettes, usually 2–6 cm long; fls (7a) 12–14 mm long excl. ovary; note lvs mainly without hairs above. EP, NL, MU, KI, SE; Vic.

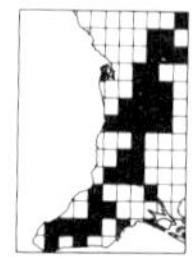

8. *Goodenia geniculata*, **bent goodenia**: Perennial herb to 10 cm tall; lvs mostly in basal rosettes, 3–10 cm long; fls (8a) 14–16 mm long excl. ovary; note lvs hairy above. EP, NL, MU, YP, KI, SE; Vic., Tas.

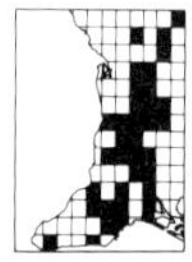

9. *Goodenia ovata*, **hop goodenia**: Shrub to 2 m tall; lvs along the stems, 3–8 cm long; fls (9a) 10–19 mm long excl. ovary; note lvs thinner than in *G. varia*. FR, EP, NL, KI, SE; all States except WA and NT.

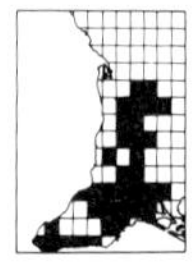

10. *Goodenia pinnatifida*, **scrambled eggs**: Perennial herb to 30 cm tall; lvs in a basal rosette, 2–10 cm long; fls (10a) 8–19 mm long excl. ovary. NW, LE, NU, GT, FR, EA, EP, NL, MU, YP, SE; all States except NT.

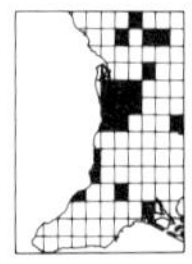

11. *Goodenia affinis*, **mallee goodenia**: Perennial herb to 8 cm tall; lvs in a basal rosette, 2–9 cm long; fls (11a) 12–15 mm long excl. ovary; note the star-shaped hairs on the upper surface of the lvs. EP, NL, MU, YP, SE; WA, NSW.

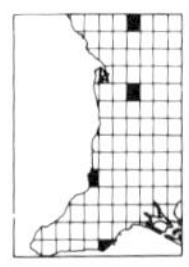

12. *Goodenia varia*, **sticky goodenia**: Shrub to 1 m tall; lvs along the stems, 2–4 cm long, glandular; fls (12a) 8–18 mm long excl. ovary; note the lvs are thicker than in *G. ovata*. NU, GT, FR, EP, NL, MU, YP, KI, SE; Vic. 3 other species.

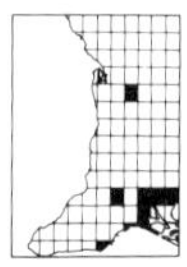

(*Continued*)

1a
1
2a
2
3a
3
4a
4b
4
5a
5
6a
6
7a
7
8a
8
9a
9
10a
10
11a
11
12a
12

GOODENIACEAE
(Continued)

1. *Scaevola angustata*, a **fanflower**: Shrub to 1 m tall; lvs 2-10 cm long, hairless; fls pale-blue to violet, 11-20 mm long excl. ovary; note lvs thinner than in *S. crassifolia* and fls larger. EP, YP, KI, SE.

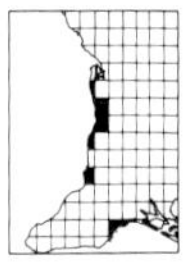

2. *Scaevola albida*, **small-fruited fan-flower**: Shrub 5-50 cm tall; lvs 6-50 mm long, hairy at least when young; fls pale-blue to white, 5-12 mm long excl. ovary. FR, NL, MU, YP, KI, SE; all States except WA and NT.

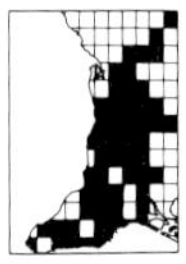

3. *Scaevola calendulacea*, **dune fanflower**: Prostrate shrub to 40 cm tall; lvs 3-8 cm long, hairy at least when young; fls bright-blue, 12-18 mm long excl. ovary. SE; all States except WA and NT. *S. aemula*, **fairy fanflower**, is a herb with similar lvs but larger fls which are sometimes white.

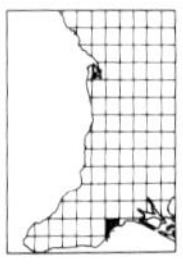

4. *Scaevola crassifolia*, **cushion fanflower**: Shrub to 1.5 m tall; lvs thick, 1.5-8.0 cm long, hairless; fls white to blue or pale purple, 8-11 mm excl. ovary. EP, YP, KI, SE; WA.

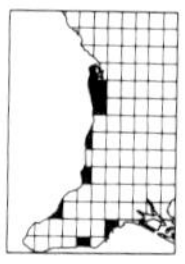

5. *Scaevola linearis*, **rough fanflower**: Shrub to 60 cm tall; lvs 2-8.5 mm long, hairy at least when young; fls blue, 12-20 mm long excl. ovary; note narrow lvs. EP, YP, KI, SE; WA.

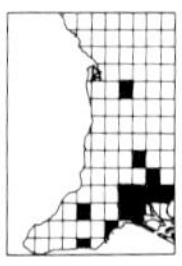

6. *Selliera radicans*, **selliera**: Prostrate perennial herb; lvs usually 1-5 cm long, succulent; fls white inside, pale-purple outside, 6-9 mm long excl. ovary. EP, MU, KI, SE; NSW, Vic., Tas.; New Zealand, S. America. Wet areas usually with high salinity.

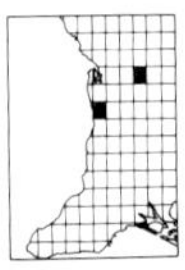

7. *Velleia arguta*, a **spur velleia**: Perennial herb 10-40 cm tall; lvs 4-12 cm long; fls (7a) 12-20 mm long excl. ovary; note the lack of hairs on the flower stalk; seed (7b) c. 4 mm diam. NW, NU, FR, EA, EP, NL, MU, YP, SE; WA, NSW, Vic.

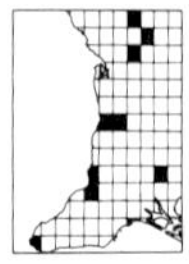

8. *Velleia paradoxa*, a **spur velleia**: Perennial herb 10-40 cm tall; lvs 7-20 cm long; fls (8a) 10-20 mm long excl. ovary; note the hairy flower stalks; seed (8b) c. 4 mm diam. NW, FR, EA, EP, NL, MU, YP, SE; Qld, NSW, Vic.

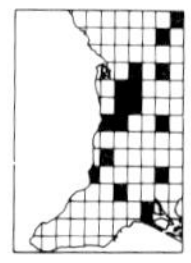

STYLIDIACEAE

9. *Levenhookia dubia*, **hairy stylewort**: Annual hairy herb to 8 cm tall; lvs 2-8 mm long; fls (9a) with whitish-pink petals 3-5 mm long. FR, EP, NL, MU, YP, KI, SE; all States except NT and Qld.

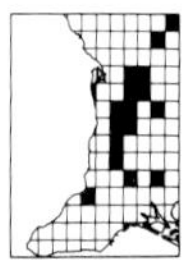

10. *Levenhookia pusilla*, **midget stylewort**: Annual hairless herb to 5 cm tall; lvs 3-6 mm long; fls (10a) with white or pinkish petals c. 2 mm long. SE; WA.

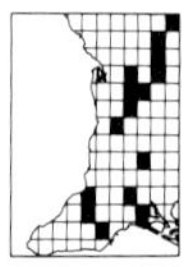

11. *Stylidium calcaratum*, **book trigger-plant**: Annual herb c. 8 cm tall; lvs 2-6 mm long; fls (11a) with white or pink petals 4-5 mm long. EP, YP, KI, SE; WA, Vic. 3 other annual species.

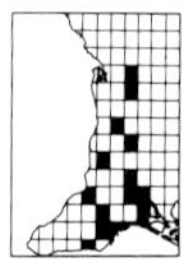

12. *Stylidium graminifolium*, **grass trigger-plant**: Perennial herb to 60 cm tall; lvs 5-30 cm long; fls (12a) with pink petals 8-11 mm long. KI, SE; all States except WA and NT.

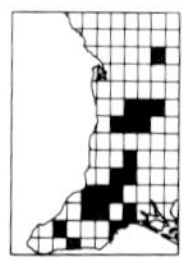

1
2
3
4
5
6
7
7a
8
8a
8b
9
9a
10
10a
11
11a
12
12a

Plate LXIII

COMPOSITAE

1. ******Chrysanthemoides monilifera*, **boneseed**: Shrub to 1.5 m tall; lvs 3-6 cm long; the yellow rays 7-14 mm long; fruit (1a) fleshy (unique for this family), 6-8 mm across. A serious weed. FR, EP, MU, YP, SE; all States except NT; native to S. Africa.

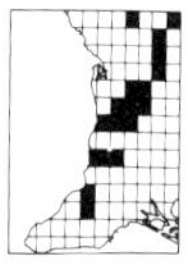

2. **Euryops abrotanifolius*, **euryops**: Shrub to 1 m tall; lvs 2-5 cm long; head (2a) with yellow rays 20-24 mm long; fruit (2b) c. 3 mm long. Vic., Tas.; native to S. Africa.

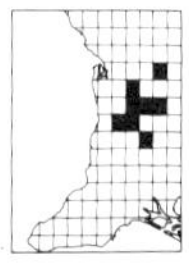

3. *Podolepis rugata*, **pleated podolepis**: Perennial herb to 60 cm tall; lvs to 8 cm long; heads (3a) with yellow rays 8-17 mm long; fruit (3b) 2-3 mm long. LE, NU, FR, EP, NL, MU, YP, KI, SE; WA, Vic.

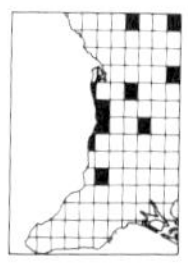

4. *Podolepis tepperi*, **delicate everlasting**: Annual herb 3-15 cm tall; lvs 5-20 mm long; heads (4a) 4-5 mm long; rays 0; outer female florets (4b) and inner bisexual florets (4c) 3.0-4.5 mm long. EP, NL, MU, YP; WA, Vic. 3 other species.

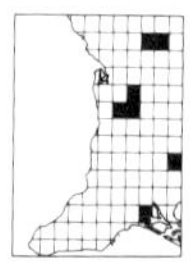

5. *Senecio glomeratus*, **annual fireweed**: Annual herb 50-120 cm tall; lvs 5-9 cm or more long; heads (5a) 4.0-5.5 mm long; rays 0; fruits (5b) 1.3-1.7 mm long, the pappus bristles falling off early. EP, SE; all States except NT and Qld; New Zealand.

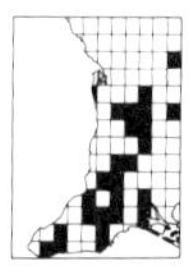

6. *Senecio glossanthus*, **slender groundsel**: Annual herb 3-40 cm tall; lvs 8-46 mm long; heads (6a) 5-6 mm long; the yellow rays 0.8-1.5 mm long; fruits (6b) c. 2-3 mm long, the pappus bristles falling off early. All regions; all mainland States.

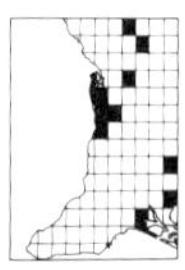

7. *Senecio hispidulus*, **rough fireweed**: Herb or shrub 30-120 cm tall; lvs 3-6 cm long; heads (7a) 5-7 mm long; rays 0; fruits (7b) 1.5-2.7 mm long, the pappus bristles falling off early. SE; all States except NT.

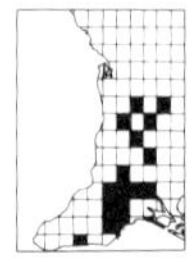

8. *Senecio hypoleucus*, **pale groundsel**: Shrub 80-120 cm tall; lf blades 6-15 mm long; heads (8a) 7-10 mm long; rays 0; florets (8b) 5-7 mm long; fruits (9b) c. 2 mm long. Vic.

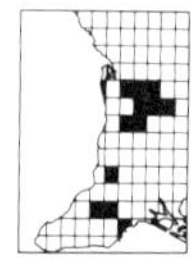

9. *Senecio lautus*, **variable groundsel**: Herb or shrub 10-80 cm tall; lvs 1-7 cm long, sometimes fleshy; heads (9a) 5-6 mm long; the yellow rays 8-12 mm long; fruits c. 2.5-3.0 mm long. All regions; all States.

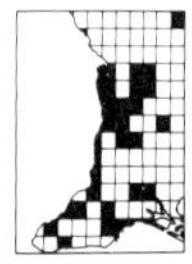

10. **Senecio pterophorus*, **African daisy**: Shrub usually 80-150 cm tall; lvs usually 5-8 cm long; heads (10a) 5-6 mm long; ray florets (10b) with rays 4-7 mm long; disk florets (10c) c. 4 mm long. EP, MU, KI; Vic.; native to S. Africa.

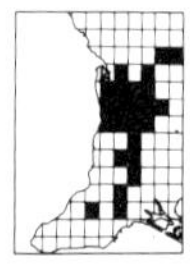

11. *Senecio quadridentatus*, **cotton fireweed**: White cottony herb 40-100 cm tall; lvs usually 5-10 cm long; heads (11a) 7.5-10 mm long; rays 0; fruits (11b) 2.5-4.0 mm long. All regions; all States; New Zealand.

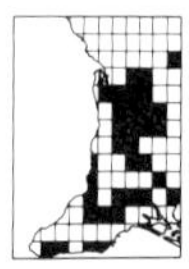

12. **Senecio vulgaris*, **common groundsel**: Annual herb 10-40 cm tall; lvs 4-8 cm long, variable in shape (12a and 12b); heads (12c) 6-7 mm long; rays 0; fruits (12d) c. 2 mm long. SE; all States; native to Europe. 8 or 9 other species.

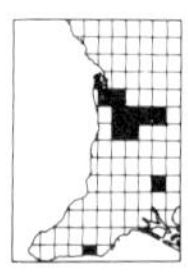

(*Continued*)

1
1a
2
2a
2b
3
3a
3b
4
4a
4c
4b
5
5a
5b
6
6a
6b
7
7a
7b
8
8a
8b
9
9a
9b
10
10a
10b
10c
11
11a
11b
12
12a
12b
12c
12d

Plate LXIV

COMPOSITAE

(Continued)

1. *Angianthus tomentosus*, **hairy cup-flower**: Cobwebby annual herb 5–35 cm tall; lvs 5–50 mm long; infl (1a) of tightly packed heads. 7–50 mm long; fruit (1b) 0.5–0.8 mm long, with a crown of scales and a few bristles. LE, NU, GT, FR, EA, EP, NL, MU, YP, SE; WA, NSW, Vic. *A. preissianus*, **common cup-flower**, has a rounder infl exceeded by its bracts.

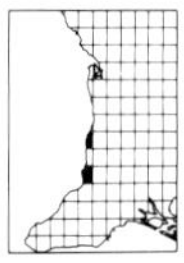

2. *Calocephalus brownii*, **cushion-bush**: Shrub to c. 1 m tall; lvs 1.5–14.0 mm long; infl (2a) of tightly packed heads, 6–16 mm long; fruit (2b) 1.2–1.9 mm long. NU, EP, YP, KI, SE; WA, Vic., Tas. 2 other species.

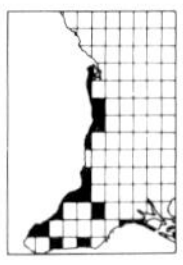

3. *Cotula australis*, **common cotula**: Herb to 12 cm tall, with weak stems; lvs 1–3 cm long; heads (3a) 4–5 mm diam.; fruit (3b) c. 1 mm long. FR, EP, NL, MU, YP, KI, SE; all States except NT; New Zealand. *C. bipinnata*, **ferny cotula**, is similar but almost entirely hairless.

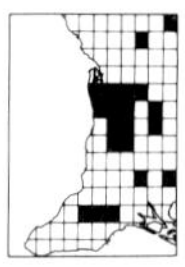

4. *_Cotula coronopifolia_, **waterbuttons**: Perennial to 60 cm tall, with weak stems; lvs 1–8 cm long; heads (4a) 8–12 mm diam.; fruit (4b) 1.5–2.0 mm long. Grows at the edge of water. LE, FR, EP, NL, MU, YP, KI, SE; all States except NT; native to S. Africa.

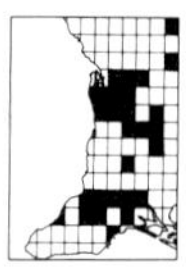

5. *Cotula vulgaris*, **slender cotula**: Annual to 20 cm tall; lvs 1–4 cm long; heads (5a) 3–5 mm diam.; fruit (5b) c. 1.5 mm long. EP, MU, YP, KI, SE; Vic., Tas.

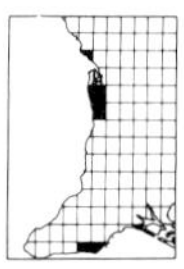

6. *Craspedia glauca*, **bachelors buttons**: Annual herb 10–80 cm tall; lvs along stems 4–12 cm long; infl (6a) of tightly packed heads, 1.5–4.0 cm diam., fruit (6b) c. 3 mm long. FR, EP, NL, MU, YP, KI, SE; all States except NT.

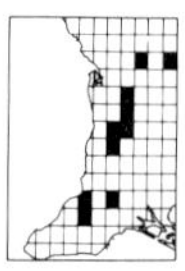

7. *Pseudognaphalium luteo-album*, **cudweed**: Herb 15–45 cm tall; lvs along stems 1.5–5.0 cm long; heads (7a) c. 4 mm long; fruit (7b) c. 0.6 mm long, the pappus bristles readily falling off. NW, LE, GT, FR, EA, EP, NL, MU, YP, KI, SE; all States; almost cosmopolitan.

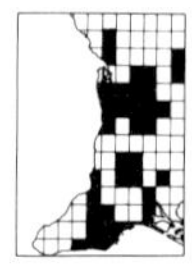

8. *Gnaphalium involucratum*, **star cudweed**: Cobwebby perennial 12–50 cm tall; lvs 3–20 cm long, all along the stem; heads (8a) 4–5 mm long; fruit (8b) c. 0.6 mm long, the pappus bristles c. 3 mm long and falling together. FR, EP, MU, KI, SE; all States; New Zealand.

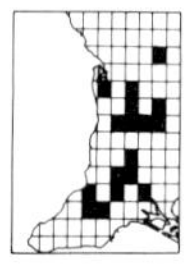

9. *_Gnaphalium spicatum_, **spiked cudweed**: Annual or biennial herb 7–40 cm tall; lvs along stems 1–6 cm long, basal lvs also present; heads (9a) 3.5–4.5 mm long; fruit (9b) c. 0.7 mm long, the pappus bristles 2.0–2.5 mm long and falling together. MU, SE; WA, NSW, Vic.; native to S. America. 4 other species.

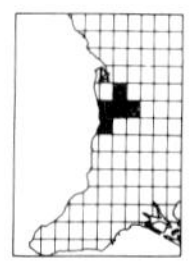

10. *Ixodia achillaeoides*, **ixodia**: Shrub 10–200 cm tall; lvs 0.5–9.0 cm long; heads (10a) with the white 1–6 mm long rays formed from the bracts; fruit (10b) 0.7–2.0 mm long, pappus bristles 0. FR, EP, NL, MU, YP, KI, SE; Vic.

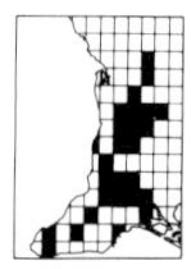

11. *Triptilodiscus pygmaeus*, **common sunray**: Annual herb 2–12 cm tall; lvs 0.5–2.0 cm long; heads (11a) 6–7 mm long; fruits (11b) c. 3 mm long, with up to 3 well formed pappus bristles. FR, EP, NL, MU, SE; all States except NT and Tas.

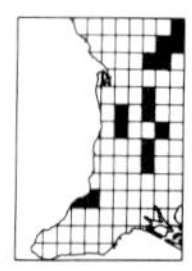

12. *_Vellereophyton dealbatum_, **white cudweed**: Usually annual 10–50 cm tall; lvs along stem 0.8–5.0 cm long; heads (12a) with the white c. 0.6 mm long rays formed from the bracts; fruits (12b) c. 0.4 mm long, the pappus bristles c. 1.5 mm long and soon falling off. MU, KI, SE; all States except NT and Qld; native to S. Africa.

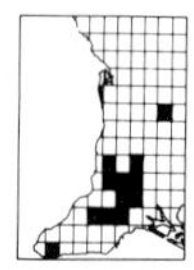

(Continued)

1a
1
1b
2a
2
2b
3b
3a
3
4a
4
4b
5
5b
5a
6a
6b
6
7a
7
7b
8a
8b
8
9
9a
9b
10
10a
10b
11
11a
11b
12a
12b
12

Plate LXV

COMPOSITAE

(*Continued*)

1. *Centipeda cunninghamii*, **common sneezeweed**: Perennial herb to 20 cm tall; lvs 1–3 cm long; heads (1a) 4–7 mm diam.; fruit (1b) 1.5–2.5 mm long; pappus 0. LE, GT, FR, EA, EP, NL, MU, KI, SE; all mainland States. *C. minima*, **spreading sneezeweed**, is an annual with stems along the ground.

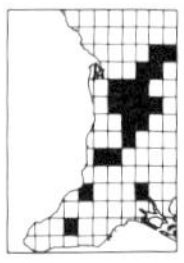

2. *Minuria leptophylla*, **Minnie daisy**: Shrub 5–40 cm tall; lvs 1–4 cm long; heads (2a) c. 15 mm diam., with white or purple 5–9 mm long rays; fruit (2b) 1.6–3.0 mm long. NW, LE, NU, GT, FR, EA, EP, NL, MU, YP, SE; all mainland States.

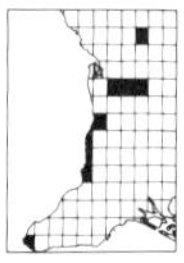

3. *Podotheca angustifolia*, **sticky long-heads**: Annual herb 4–20 cm tall; lvs along stem 2–6 cm long; heads (3a) 2.5–4.0 cm long, sticky; fruit (3b) 2–3 mm long, with very long pappus bristles. NU, GT, FR, EP, MU, YP, KI, SE; all States except NT and Qld.

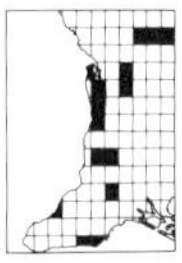

4. *Pogonolepis muelleriana*, **stiff cup-flower**: Flat annual herb to 9 cm tall; lvs 2–4 mm long; infl (4a) 4–5 mm long, made up of many smaller heads; fruit (4b) c. 1 mm long, pappus 0. FR, EP, NL, MU, YP, KI; NSW, Vic.

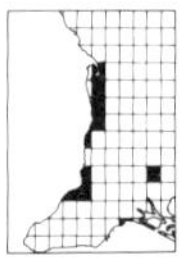

5. *Quinetia urvillei*, **quinetia**: Annual herb 3–12 cm tall; lvs 7–15 mm long; heads (5a) c. 5 mm long; fruit (5b) c. 4 mm long, with long tapering pappus scales. EP, NL, MU, SE; WA, Vic.

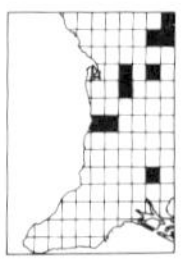

6. *Soliva pterosperma*, **jo jo**: Annual flat on the ground 3–10 cm long; lvs 1.5–3.0 cm long; heads (6a) c. 5 mm diam.; fruit (6b) 3–4 mm long, with characteristic wings. NL; WA, NSW, Vic.; native to S. America.

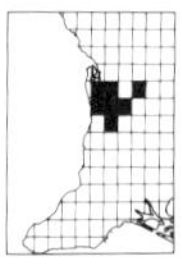

7. *Toxanthes muelleri*, **common bow-flower**: Annual herb 1–3 cm tall; lvs 0.5–2.0 cm long; heads (7a) 2.5–3.5 mm long; fruit (7b) 4–5 mm long; pappus 0. NW, NU, FR, EP, NL, MU, YP, KI, SE; Vic.

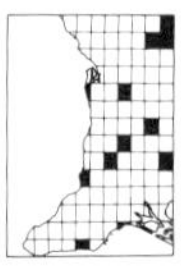

8. *Vittadinia australasica*, a **vittadinia**: Annual or perennial herb 15–25 cm tall; lvs 1.0–2.5 cm long; heads (8a) 5.5–7.5 cm long, with white to mauve rays; fruit (8b) 3.0–4.5 mm long, with longer pappus bristles. NU, EP, MU, YP, KI, SE; all States except NT and Qld.

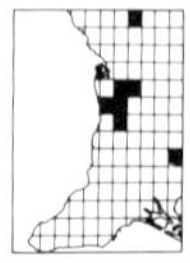

9. *Vittadinia blackii*, **western New Holland daisy**: Shrub 10–30 cm tall; lvs cylindrical, 1–2 cm long; heads (9a) 7–9 mm long, with violet rays; fruit (9b) 4.5–5.5 mm long, with pappus bristles about as long as the fruit. LE, FR, EA, EP, NL, MU, YP, SE; WA.

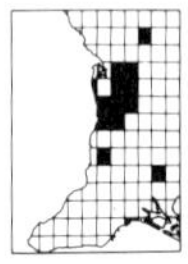

10. *Vittadinia cuneata*, **fuzzweed**: Annual or perennial herb 10–40 cm tall; lvs 1.0–2.5 cm long; heads (10a) 5–8 mm long, with pale-blue to mauve rays; fruit (10b) 4–7 mm long, the pappus bristles about as long as the fruit. FR, EA, EP, NL, MU, YP, KI, SE; all States except WA and NT.

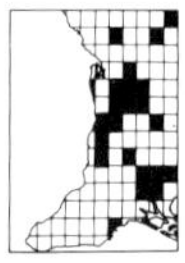

11. *Vittadinia gracilis*, **woolly New Holland daisy**: Shrub 10–40 cm tall, with dense white woolly-hairy stems; lvs 1–4 cm long; heads (11a) 7–8 mm long, with purple rays; fruit (11b) 4.5–6.0 mm long, with pappus hairs about as long as the fruit; note the glandular as well as plain hairs on the fruit. NW, FR, EA, EP, NL, MU, YP, KI, SE; all States except NT and Qld.

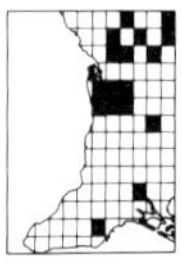

12. *Vittadinia megacephala*, **giant New Holland daisy**: Perennial herb 5–25 cm tall; lvs 1.5–4.5 cm long; heads (12a) 10–14 mm long, with dark purple to blue rays; fruit (12b) 6–8 mm long, the pappus bristles about as long as the fruit; note the absence of ridges from fruit. EP, NL, MU, YP, SE; Vic., Tas. 3 other species.

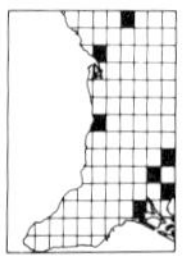

(*Continued*)

1b
1a
1
b
2b
2a
2
3b
3
3a
4
4a
5a
5
5b
6a
6b
6
7a
7b
7
8a
8
8b
9a
9
9b
10b
10
10a
11
11b
11a
12b
12
12a

Plate LXVI

COMPOSITAE
(Continued)

1. ******Aster subulatus*, **wild aster**: Perennial herb to 1.8 m tall; lvs 1–12 cm long; heads (1a) 5–7 mm long, with pale-mauve rays; fruit (1b) c. 2 mm long. LE, FR, EA, NL, MU, SE; all States except Qld; native to N. America.

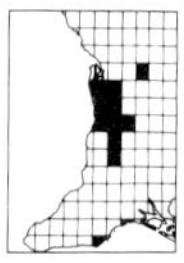

2. *Brachycome ciliaris*, **variable daisy**: Perennial herb to 45 cm tall; lvs 1–6 cm long; heads (2a) with white to mauve 2–7 mm long rays; fruit of 2 kinds c. 1.5–2.5 mm long, only the inner ones (2b) winged. NW, LE, NU, GT, FR, EA, EP, NL, MU, YP, SE; all States.

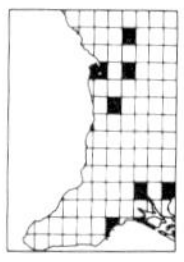

3. *Brachycome diversifolia*, **large-headed daisy**: Perennial herb to 45 cm tall; lvs 2–9 cm long; heads (3a) with white rays 10–12 mm long; fruit (3b) 2–3 mm long. SE; all States except WA.

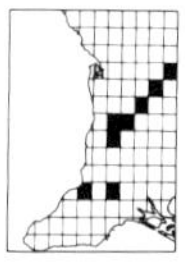

4. *Brachycome lineariloba*, **hard-headed daisy**: Annual herb to 15 cm tall; lvs 3–8 cm long; heads (4a) with white rays 1–6 mm long; fruit (4b) 3.0–4.5 mm long. LE, NU, GT, FR, EA, EP, NL, MU, YP, KI, SE; all States except Qld and Tas.

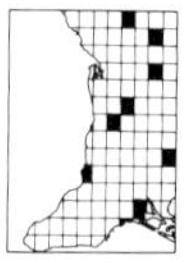

5. *Brachycome parvula*, **coast daisy**: Annual or perennial herb to 40 cm tall; lvs 2–10 cm long, mainly along stems; heads (5a) with white to lilac rays 3–5 mm long; fruit (5b) 1.5–2.0 mm long. EP, MU, SE; Vic., Tas.

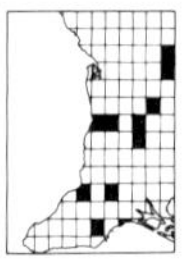

6. *Brachycome perpusilla*, **tiny daisy**: Annual herb to 10 cm tall; lvs 0.4–3.0 cm long; heads (6a) with inconspicuous white rays c. 1 mm long; fruit (6b) c. 2 mm long; note fringed wings to fruit. FR, EP, NL, MU, YP, SE; WA, NSW, Vic. 8 or 9 other species.

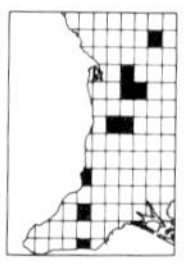

7. **Conyza albida*, **tall fleabane**: Annual herb to 2 m tall; lvs along stems 3–6 cm long; heads (7a) 4–5 mm long; fruit (7b) c. 1 mm long. MU, SE; all States except NT and Qld; native to S. America.

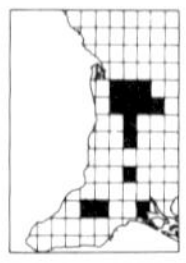

8. **Conyza bonariensis*, **flaxleaf fleabane**: Annual herb to 1 m tall; lvs along stems 2–6 cm long; heads (8a) 5–6 mm long; fruit (8b) c. 1 mm long. NW, GT, FR, EA, EP, NL, MU, YP, SE; all States; native to S. America. 2 other species.

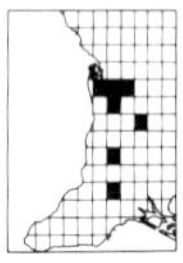

9. *Leptorhynchos squamatus*, **scaly buttons**: Perennial herb 10–40 cm tall; lvs mostly 1–3 cm long; heads (9a) 7–12 mm long, with long hairs on the bracts; fruit (9b) 2.5–3.0 mm long. FR, EP, NL, MU, YP, SE; NSW, Vic., Tas. 3 other species.

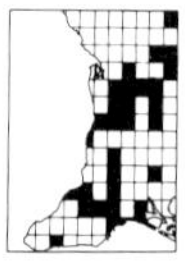

10. **Logfia gallica*, **narrow cudweed**: Annual herb to 20 cm tall; lvs 6–20 mm long; heads (10a) 2.5–3.5 mm long; fruit (10b) c. 0.6 mm long, with pappus bristles to 2 mm long. LE, MU, SE; WA, Vic.; native to Europe.

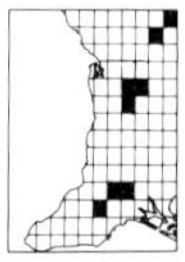

11. *Millotia myosotidifolia*, **broad-leaved millotia**: Annual herb to 20 cm tall; lvs 1–5 cm long; heads (11a) 4–8 mm long; fruit (11b) 4–8 mm long. NU, GT, FR, EA, EP, NL, MU, YP, KI, SE; all States except NT and Qld.

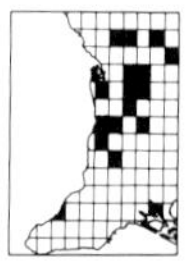

12. *Millotia tenuifolia*, **soft millotia**: Annual herb with glandular hairs, to 12 cm tall; lvs 0.7–2.2 cm long; heads (12a) 4–7 mm long; fruit (12b) 4–7 mm long. FR, EP, NL, MU, YP, KI, SE; all States except NT and Qld.

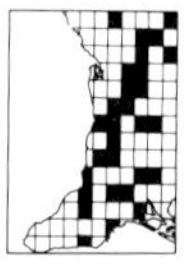

(Continued)

1a
1
1b
2a
2c
2d
2b
2
3b
3a
3
b
4a
4
5a
5b
5
6a
6b
6
7b
7a
7
8a
8b
8
9a
9b
9
10b
10a
10
11b
11a
11
12b
12a
12

Plate LXVII

COMPOSITAE

(*Continued*)

1. *Cassinia uncata*, **sticky cassinia**: Shrub 20–150 cm tall; lvs 1.0–4.5 cm long; infl. of numerous densely arranged 3–5 mm long heads (1a); fruit c. 0.8 mm long, with pappus bristles about twice as long. FR, EP, NL, MU, YP, KI, SE; NSW, Vic., Tas.

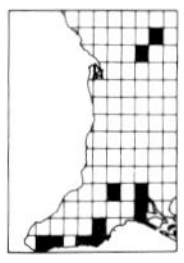

2. *Olearia axillaris*, **coast daisy-bush**: Shrub 2–3 m tall; lvs 1.0–3.5 cm long; heads (2a) 4–5 mm long, without rays; fruit (2b) c. 1.5 mm long. EP, NL, YP, KI, SE; all States except NT and Qld.

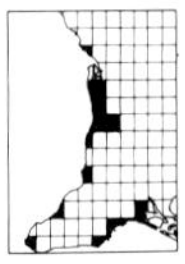

3. *Olearia ciliata*, **fringed daisy-bush**: Shrub 15–30 cm tall; lvs 0.5–2.0 cm long; heads (3a) 7–8 mm long, with lilac to blue rays 12–15 mm long; fruit (3b) c. 2 mm long. FR, EA, EP, NL, MU, YP, KI, SE; all States except NT and Qld.

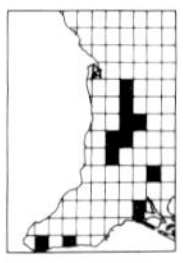

4. *Olearia floribunda*, **heath daisy-bush**: Shrub 1–2 m tall; lvs 0.8–2.5 mm long; heads (4a) 3–4 mm long, with white to pale pink rays 3–5 mm long; fruit (4b) 1.5–2.0 mm long. EP, NL, MU, YP, SE; all States except NT and Qld.

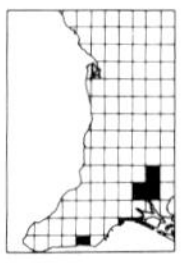

5. *Olearia glandulosa*, **swamp daisy-bush**: Shrub, sticky, to 1.5 m tall; lvs 1–6 cm long; heads (5a) 2–3 mm long, with white to pink rays 3–6 mm long; fruit (5b) c. 1 mm long. SE; NSW; Vic., Tas.

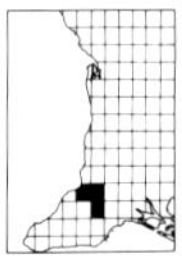

6. *Olearia grandiflora*, **Mount Lofty daisy-bush**: Shrub 30–90 cm tall; lvs 4–10 cm long; heads (6a) 16–24 mm long, with white rays 2–3 cm long; fruit (6b) 5–6 mm long. Confined to this region, in wooded places.

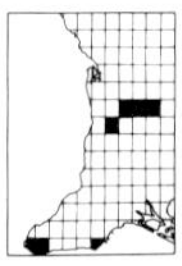

7. *Olearia lanuginosa*, **woolly daisy-bush**: Shrub to 1.5 m tall; lvs 1.0–2.5 mm long; heads (7a) 4–5 mm long, with white rays 3–5 mm long; fruit (7b) c. 1.5 mm long. EP, NL, MU, YP, SE; NSW, Vic.

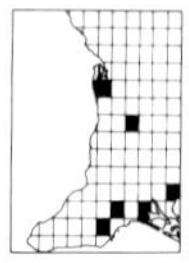

8. *Olearia pannosa*, **silver-leaved daisy**: Shrub to 1.5 m tall; lvs 3–9 cm long; heads (8a) 15–22 mm long, with white to pale mauve rays 2–3 cm long; fruit (8b) 5–8 mm long. FR, EP, NL, MU, YP, SE; NSW, Vic.

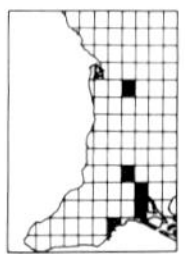

9. *Olearia passerinoides*, a **daisy-bush**: Shrub to 2.5 m tall; lvs 4–20 mm long; heads (9a) 5–6 mm long, with white to pale pink rays 4–10 mm long; fruit (9b) 1.5–2.0 mm long. EP, NL, MU, YP; WA, Vic.

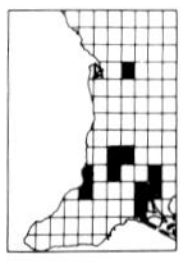

10. *Olearia ramulosa*, **twiggy daisy-bush**: Shrub to 1.5 m tall; lvs usually 4–12 mm long; heads (10a) 3–6 mm long, with white, pink or pale blue rays 2–5 mm long; fruit 1–2 mm long. GT, FR, EP, NL, MU, YP, KI, SE; all States except NT

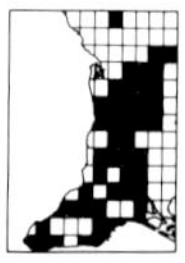

11. *Olearia teretifolia*, **cypress daisy-bush**: Shrub to 2 m tall; lvs 2–5 mm long; heads (11a) c. 4 mm long, with white rays 5–6 mm long; fruit (11b) c. 1.5 mm long. EP, NL, MU, KI; NSW, Vic.

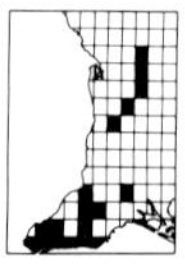

12. *Olearia tubuliflora*, **rayless daisy-bush**: Shrub to 2 m tall; lvs 5–10 mm long; heads (12a) c. 3 mm long, without rays; fruit (12b) c. 1 mm long. FR, EP, NL, SE; NSW, Vic. 3 other species.

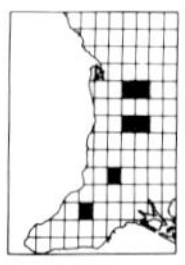

(*Continued*)

1
1a
2b
2
2a
3a
3
3b
4a
4b
4
5
5b
5a
6
6b
6a
7a
7b
7
8
8a
8b
9a
9b
9
10b
10
10a
11a
11
11b
12
12a
12b

PLATE LXVIII

COMPOSITAE
(Continued)

1. *Helichrysum apiculatum*, **common everlasting**: Perennial herb 7–60 cm tall; lvs 1–6 cm long; heads (1a) 7–15 mm diam., with golden-yellow or brownish or rarely whitish or pinkish rays formed from the bracts; fruit (1b) c. 1 mm long. All regions; all States.

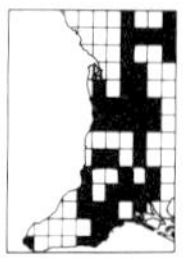

2. *Helichrysum baxteri*, **fringed everlasting**: Perennial herb 15–40 cm tall; lvs 0.5–3.0 cm long; heads (2a) 2–3 cm diam., with white rays formed from the bracts; fruit (2b and 2c) c. 1 mm long. ?FR, NL, MU, YP, SE; NSW, Vic.

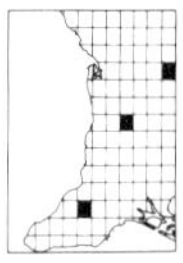

3. *Helichrysum blandowskianum*, **woolly everlasting**: Perennial herb 20–50 cm tall; lvs usually 2–4 cm long; heads (3a) 2.0–2.5 cm diam., with white or pink rays formed from the bracts; fruit (3b) c. 1 mm long. FR, MU, SE; Vic.

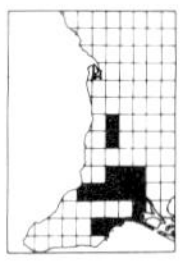

4. *Helichrysum bracteatum*, **golden everlasting**: Herb 20–80 cm tall; lvs mostly 1.5–10.0 cm long; heads (4a) 25–50 mm diam., with bright yellow or white rays formed from the bracts; fruit (4b) 2–3 mm long. NW, LE, NU, FR, EA, EP, NL, MU, YP, KI; all States. Commonly grown as a garden plant.

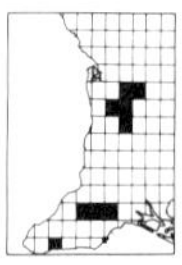

5. *Helichrysum leucopsidium*, **satin everlasting**: Perennial herb 15–50 cm tall; lvs mostly 2–5 cm long; heads (5a) 25–35 mm diam., with white rays formed from the bracts; fruit (5b) c. 2 mm long. NU, FR, EP, NL, MU, YP, KI, SE; all States except NT and Qld.

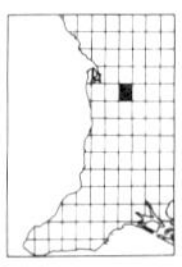

6. *Helichrysum obtusifolium*, an **everlasting**: Perennial 10–30 cm tall; lvs mostly 0.4–2.5 cm long; heads (6a) 2–3 cm diam., with white rays formed from the bracts; fruit (6b) to 1 mm long. EP, MU, YP, KI, SE; all States except NT and Qld.

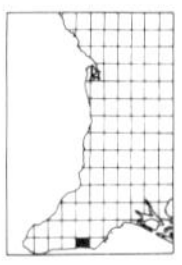

7. *Helichrysum paralium*, **coast everlasting**: Shrub 1–2 m tall; lvs 1.0–2.5 cm long; heads (7a) 5–7 mm diam., with white rays formed from the bracts; fruit (7b) to 1 mm long. Coastal dunes. SE; Vic., Tas.

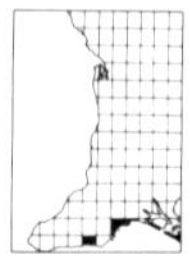

8. *Helichrysum scorpioides*, **button everlasting**: Perennial herb 20–50 cm tall; lvs mostly 4.5–9.0 cm long; heads (8a) 2–3 cm diam., with yellow to pale yellow rays formed from the bracts; fruit (8b) c. 1.5 mm long. YP, SE; NSW, Vic., Tas.

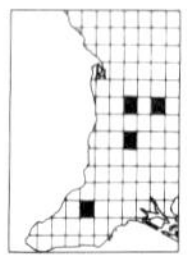

9. *Helichrysum semipapposum*, **clustered everlasting**: Perennial herb, aromatic, 15–60 cm tall; lvs 2–50 mm long; heads (9a) 6–7 mm diam., with yellow rays formed from the bracts; fruit (9b) c. 1 mm long. NW, NU, FR, EA, EP, NL, MU, YP, KI, SE; all States. 3 other species.

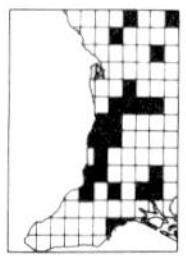

10. *Helipterum corymbiflorum*, **grey sunray**: Annual herb 8–30 cm tall; lvs 1–4 cm long; heads (10a) 12–16 mm long, with white rays formed from the bracts; fruit (10b) c. 2.5 mm long. LE, GT, FR, EA, EP, NL, MU, YP, SE; NT, Qld, NSW, Vic.

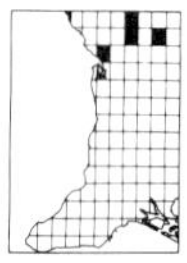

11. *Helipterum demissum*, **dwarf sunray**: Annual herb forming moss-like clumps to 2.5 cm high; lvs 2–5 mm long; heads (11a) c. 3 mm long; fruit (11b) 0.8–1.0 mm long, with 9–12 feathery bristles which fall off early. GT, FR, EP, NL, MU, YP, KI, SE; all States except NT and Qld. 6 other species.

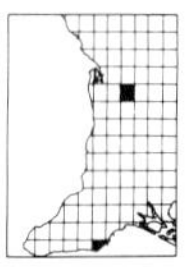

12. *Ixiolaena supina*, a **stalked ixiolaena**: Perennial herb with branches usually under 15 cm long, mainly along the ground; lvs 0.7–3.0 cm long; heads (12a) 15–23 mm diam.; fruit (12b) 1.5–2.0 mm long. EP, YP, KI; Tas. *I. leptolepis*, a **stalked ixiolaena**, is an erect bush.

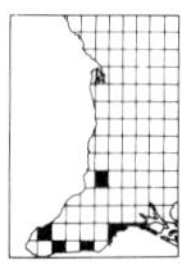

(Continued)

1a
1
1b
2
2a
2b
2c
3a
3b
3
4b
4a
4
5a
5b
5
6a
6b
6
7b
7a
7
8a
8b
8
9
9a
9b
10
10a
10b
11a
11b
11
12a
12
12b

Plate LXIX

COMPOSITAE
(Continued)

1. **Arctotheca calendula*, **Capeweed**: Herb to 30 cm high, stems absent or along the ground; lvs (1a) 5-25 cm long; heads (1b) 2.5-5.0 cm diam. NW, LE, GT, FR, EP, NL, MU, YP, KI, SE; all States; native to S. Africa.

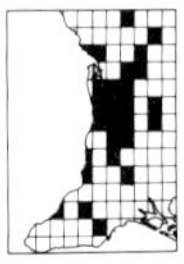

2. *Cymbonotus preissianus*, **austral bears-ear**: Stemless herb; lvs 4-15 cm long; heads (2a) 7-10 mm long, with yellow rays 6-10 mm long; fruits (2b, 2 views) 3.0-3.5 mm long, pappus 0. FR, EP, NL, MU, YP, KI, SE; WA, NSW, Vic.

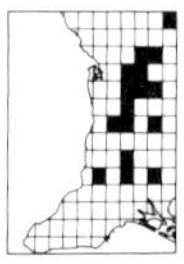

3. *Lagenifera huegelii*, **coarse bottle-daisy**: Stemless herbs; lvs 4-11 cm long; heads (3a) 8-11 mm diam., with white to blue rays 1-2 mm long; fruit (3b) 2-3 mm long. FR, EP, NL, YP, KI, SE; WA, Vic., Tas.

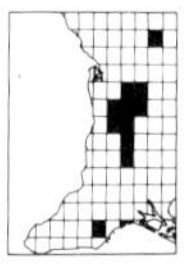

4. *Lagenifera stipitata*, **common lagenophora**: Stemless herbs; lvs 2-6 cm long; heads (4a) 5-8 mm diam., with white to blue or violet rays 2.5-4.0 mm long; fruit (4b) 3-4 mm long. KI, SE; all States except NT. *L. gracilis*, **slender lagenophora** resembles *L. huegelii* in lacking stolons and in having short rays but has only 2-4 pairs of teeth on the lvs.

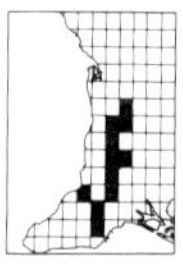

5. **Sonchus asper*, **prickly sow-thistle**: Annual or biennial herb with milky sap, 20-150 cm tall; lvs 4-20 cm long; heads (5a) 10-13 mm long; fruit (5b) 2-3 mm long. All regions; all States; native to the Northern Hemisphere.

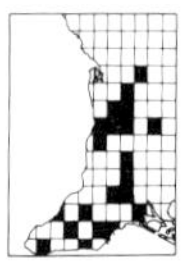

6. *Sonchus hydrophilus*, **native sow-thistle**: Annual or possibly perennial herb with milky sap, 50-120 cm tall; lvs 5-30 cm long; heads (6a) 10-12 mm long; fruit (6b) 2.7-3.7 mm long. NW, LE, FR, EP, MU, YP, KI, SE; all mainland States except NT.

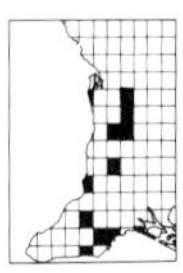

7. *Sonchus megalocarpus*, **dune thistle**: Perennial herb with rhizome and milky sap, 15-60 cm tall; heads (7a) 12-20 mm long; fruit (7b) 4.5-7.5 mm long. On coastal dunes. EP, YP, KI, SE; all States except NT.

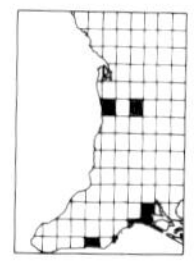

8. **Sonchus oleraceus*, **common sow-thistle**: Annual herb with milky sap, 20-110 cm tall; heads (8a) 10-12 mm long; fruit (8b) wrinkled, 2.5-3.7 mm long. All regions; all States; probably native to Europe and western Asia.

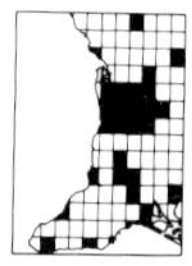

9. **Taraxacum officinalis*, a **dandelion**: Perennial herb with milky sap, 5-40 cm tall; lvs basal, 5-30 cm long; heads (9a) 2.0-4.5 cm diam.; fruit (9b) 2.5-3.5 mm long. EP, NL, MU, YP, KI, SE; all States except NT; native to Europe. *T. erythrospermum*, a dandelion, has red-brown fruits.

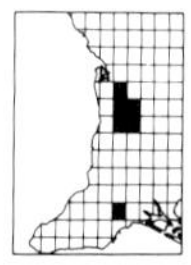

10. **Tolpis barbata*, **tolpis**: Annual herb with milky sap, 12-50 cm tall; lvs usually 2-8 cm long; heads (10a) c. 15 mm diam.; fruit (10b) c. 1.5 mm long. All States except NT and Qld; native to southern Europe.

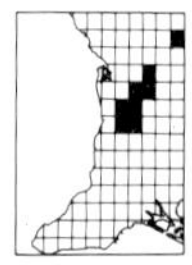

11. **Tragopogon porrifolius*, **salsify**: Biennial herb with milky sap, to 1.2 m tall; lvs along stems 3-25 cm long; heads (11a) 2.5-6.0 cm long; fruit (11b) 1.5-2.0 cm long excl. the stalk below the pappus bristles. FR, NL, MU, SE; all States except NT; native to the Mediterranean.

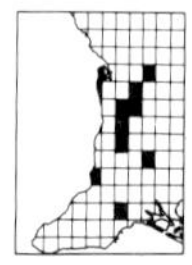

12. **Urospermum picroides*, **false hawkbit**: Annual herb with milky sap, 10-50 cm tall; lvs at base 3-16 cm long; heads (12a) 13-20 mm long; fruit (12b) 10-12 mm long. FR, EP, NL, MU, YP, KI, SE; WA, Vic.; native to the Mediterranean and S.W. Asia.

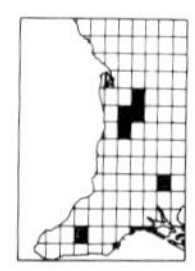

(Continued)

1b
1
1a
2a
2b
2
3b
3a
3
4a
4
4b
5a
5
5b
6b
6
6a
7b
7a
7
8a
8
8b
9b
9a
9
10a
10
10b
11a
11
11b
12a
12b
12

PLATE LXX

COMPOSITAE

(Continued)

1. ***Cichorium intybus*, chickory**: Perennial herb with milky sap, to 1 m tall; lvs at base 9–30 cm long; heads (1a) 9–12 mm long; fruit 2–3 mm long, with a crown of minute scales. NL, MU, KI, SE; all States except NT; native to Europe.

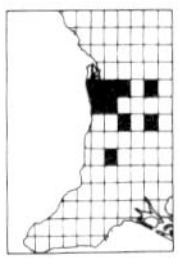

2. ***Crepis capillaris*, smooth hawksbeard**: Annual or biennial herb with milky sap, 4–90 cm tall; lvs at base 5–25 cm long; heads (2a) 5–8 mm long; fruit (2b) 1.5–2.5 mm long, without a stem below the pappus bristles, SE; all States except NT and Qld; native to Europe. *C. dioscoridis*, a **hawksbeard**, has greyer bracts on the heads and longer fruits.

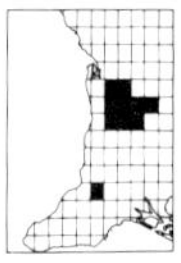

3. ***Crepis foetida*, stinking hawksbeard**: Annual herb with milky sap, 10–50 cm tall; lvs at base 4–13 cm long; heads (3a) 9–12 mm long; fruit with (3b) or without (3c) a stalk below the pappus bristles, 6–9 mm long excl. the stalk. NL, YP; WA, NSW, Vic.; native to Europe and S.W. Asia.

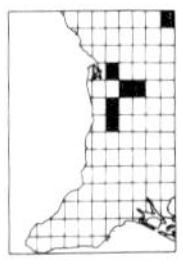

4. ***Hedypnois rhagadioloides*, Cretan weed**: Annual herb with milky sap, 3–30 cm tall; lvs at base 2–16 cm long; heads (4a) c. 1 cm diam.; fruiting head (4b); fruits (4c) 3.5–7 mm long, some with long pappus bristles. LE, NU, GT, FR, EP, NL, MU, YP, KI, SE; all States; native to Europe.

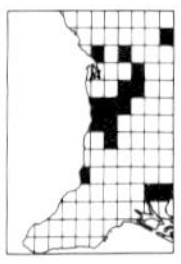

5. ***Hypochoeris glabra*, smooth catsear**: Annual herb with milky sap, 10–40 cm tall; lvs 1–20 cm long; heads (5a) 4–16 mm diam.; fruit with (5b) or without (5c) a stalk below the pappus bristles, 2.5–12.0 mm long incl. the stalk. GT, FR, EA, EP, NL, MU, YP, KI, SE; all States; native to Europe.

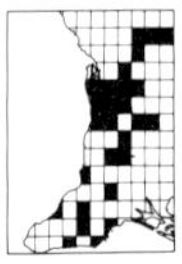

6. ***Hypochoeris radicata*, deep-rooted catsear**: Perennial herb with milky sap, 15–80 cm tall; lvs 3–24 cm long; heads (6a) 20–30 mm diam.; fruit usually with a long stalk (6b) below the pappus bristles, 8–17 mm long. EP, NL, MU, YP, KI, SE; all States except NT and Qld; native to Europe.

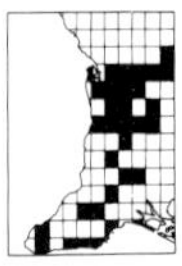

7. ***Lactuca serriola*, prickly lettuce**: Biennial herb with milky sap, 1–2 m tall; lvs 3–18 cm long; heads (7a) 6–9 mm long; fruit (7b) 6–8 mm long incl. the long stalk below the pappus bristles. FR, EA, EP, NL, MU, YP, SE; all States; native to Europe. *L. saligna*, **willow lettuce**, has narrower lvs with narrower lobes.

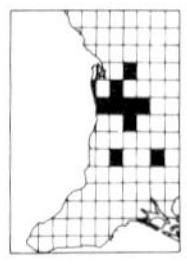

8. ***Leontodon taraxacoides*, lesser hawkbit**: Perennial herb with milky sap, 15–35 cm tall; lvs 3–17 cm long; heads (8a) 7–11 mm long; fruit with (8b) or without (8c) pappus bristles, c. 4–5 mm long. KI, SE; all States except NT and Qld; native to Europe.

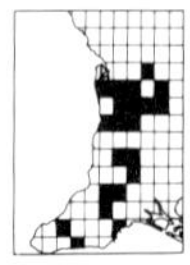

9. ***Microseris lanceolata*, yam daisy**: Perennial herb with milky sap, 10–40 cm tall; lvs 4–30 cm long; heads (9a) 16–24 mm long; fruit (9b) 6–9 mm long. GT, FR, EA, EP, NL, MU, YP, KI, SE; all States except NT and Qld.

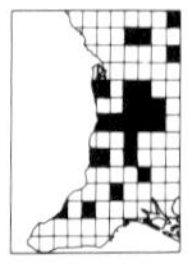

10. ***Picris echioides*, ox-tongue**: Annual or biennial herb with milky sap and characteristic hairs (10a), 30–100 cm tall; lvs at base 3–20 cm long; heads (10b) 10–15 mm long; fruit (10c) c. 3 mm long excl. the stalk below the pappus bristles. EP, MU, SE; NSW, Vic., Tas.; native to Europe.

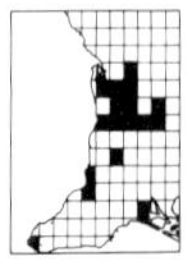

11. *Picris squarrosa*: Perennial herb with milky sap, 30–80 cm tall; lvs at base 6–22 cm long; heads (11a) 11–15 mm long; fruit (11b) 7–9 mm long. MU, YP, KI, SE; WA, Vic. **P. hieracioides*, **hawkweed picris**, has less spiny hairs on the stems.

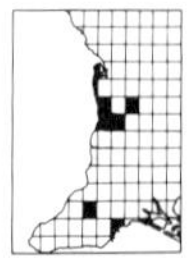

12. ***Reichardia tingitana*, false sow-thistle**: Herb 10–50 cm tall; lvs at base 2–17 cm long; heads (12a) 10–15 mm long; fruit (12b) 1.5–2.8 mm long. GT, EA, EP, NL, MU, YP, SE; WA, NSW, Vic.; native to the Mediterranean.

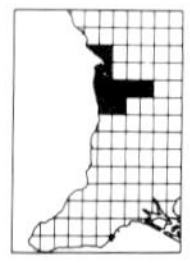

(Continued)

1a
1
2a
2
2b
3
3a
3c
3b
4b
4a
4c
4
5c
5b
5a
5
6b
6
6a
7a
7
7b
8c
8b
8a
8
9a
9
9b
10b
10
10c
10a
11
11a
11b
12b
12a
12

Plate LXXI

COMPOSITAE

(Continued)

1. **Ambrosia psilostachya*, **perennial ragweed:** Perennial herb to 2 m tall; lvs 5–12 cm long; male heads (1a) c. 3 mm diam.; female heads (1b) in small groups surrounded by leaf-like bracts, 3–4 mm long; fruiting head (1c) 3–4 mm long. MU; WA, NSW, Vic.; native to N. America. **A. tenuifolia*, **lacy ambrosia**, has more divided lvs.

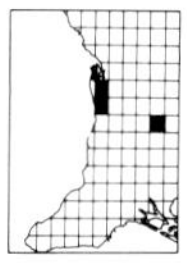

2. **Calendula arvensis*, **field marigold**: Annual herb 5–40 cm tall; lvs 2–6 cm long; heads (2a) with yellow rays 7–10 mm long; fruit (2b) of several forms, to c. 1.5 mm long. LE, FR, EA, EP, NL, MU; all States except WA and NT; native to Europe. **C. officinalis*, **garden marigold**, has larger heads.

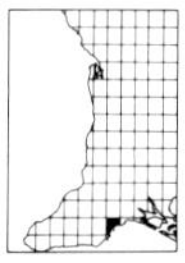

3. **Carduus tenuiflorus*, **slender thistle**: Annual herb 50–150 cm tall; lvs along stems 5–15 cm long; heads (3a) 0.6–1.0 cm diam.; fruit (3b) 4–5 mm long. FR, EP, NL, MU, YP, KI, SE; all States except NT and Qld; native to Europe. **C. pycnocephalus*, **shore thistle**, has shortly stalked heads.

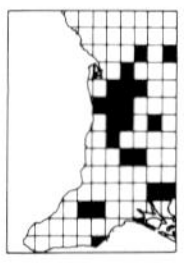

4. **Centaurea calcitrapa*, **star thistle**: Annual or biennial herb to 100 cm tall; lvs along stems 1.5–8.0 cm long; heads (4a) 8–12 mm diam.; fruit (4b) c. 3 mm long, pappus 0. NL, MU, YP, SE; all States except NT and Qld; native to Europe.

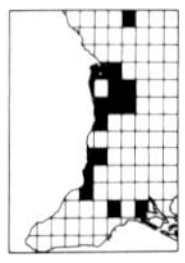

5. **Centaurea melitensis*, **Maltese cockspur**: Annual herb 20–100 cm tall; lvs along stems 1–7 cm long; heads (5a) 8–12 mm diam.; fruit (5b) c. 2.5 mm long. LE, NU, GT, FR, EA, EP, NL, MU, YP, KI, SE; all States except NT; native to Europe. 5 other species.

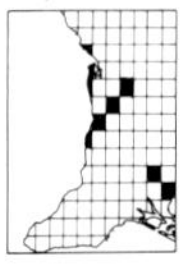

6. **Carthamus lanatus*, **saffron thistle**: Annual herb to 1 m tall; lvs along stem 3–6 cm long; heads (6a) 2.5–3.0 cm long; fruit (6b) 5–6 mm long. LE, FR, EA, EP, NL, MU, YP, SE; all States; native to Europe and Asia.

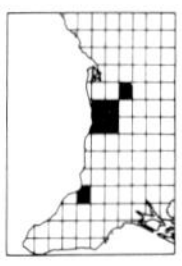

7. **Cirsium vulgare*, **spear thistle**: Biennial herb to 1.5 m tall; lvs along stems 4–25 cm long; heads (7a) 2–4 cm long; fruit (7b) 3.5–5.0 mm long; note feathery pappus. FR, EA, EP, NL, MU, YP, SE; all States; native to Europe, Asia and Africa.

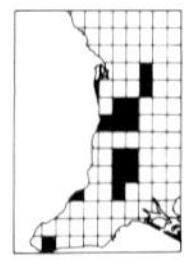

8. **Cynara cardunculus*, **artichoke thistle**: Perennial herb to 1.5 m tall; lvs along stems to 75 cm long; heads (8a) 4.5–6.0 cm diam.; fruit (8b) 6–8 mm long. EA, EP, NL, YP, SE; all States except NT; native to the Mediterranean.

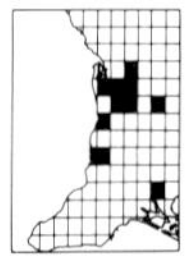

9. **Galinsoga parviflora*, **yellow weed**: Annual herb 25–50 cm tall; lvs 2–6 cm long; heads (9a) 4–7 mm diam.; inner fruits (9b) c. 1.8 mm long, the outer fruits with a pappus of small bristles. All States except NT; native to S. America.

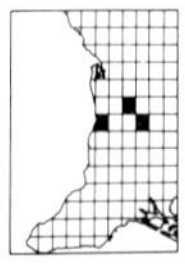

10. **Sigesbeckia orientalis*, an **Indian weed**: Annual herb to 80 cm tall; lvs 3–12 cm long; heads (10a) c. 1 cm diam.; fruit (10b) 2–3 mm long; pappus 0. KI, SE; all States; native to Europe. **S. microcephala*, **pale Indian weed**, lacks the glands on the heads.

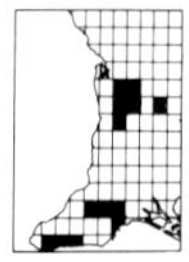

11. *Rutidosis multiflora*, **small wrinklewort**: Annual herb 1.5–5.0 cm tall; lvs mostly 3–8 mm long; heads (11a) 2–4 mm diam.; fruit (11b) c. 1 mm long excl. the pappus scales. EA, EP, NL, MU, YP, KI, SE; all States except NT and Qld.

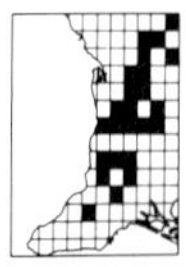

12. **Xanthium spinosum*, **Bathurst burr**: Annual herb to 1 m tall; lvs 3–8 cm long; female heads (12a) in fruit 10–12 mm long. LE, GT, FR, EA, EP, NL, MU, YP, SE; all States; probably native to tropical America. **X. californicum*, **California burr**, lacks spines on the stems.

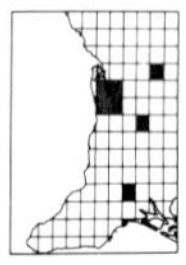

1b
1
1a
1c
2
2a
2b
3b
3a
3
4b
4a
4
5a
5b
5
6b
6a
6
7
7a
7b
8a
8
9a
9b
9
10b
10
10a
11
11a
11b
12
12a

PLATE LXXII

LILIACEAE

1. *Dichopogon fimbriatus*, **nodding chocolate-lily**: Usually 40–80 cm tall; fls (1a); 'petals' blue, violet or rarely white, 10–12 mm long; several fls from each bract. FR, EP, NL, MU, YP, KI, SE; all States except NT and Tas.

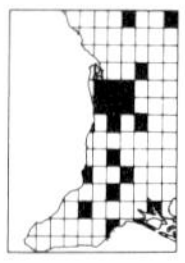

2. *Dichopogon strictus*, a **chocolate lily**: 20–100 cm tall; fls (2a); 'petals' blue, violet or rarely white, 10–12 mm long; 1 fl from each bract. NW, FR, EP, NL, MU, YP, KI, SE; all States. *Arthropodium minus*, **small vanilla-lily**, has bearded filaments below the anthers.

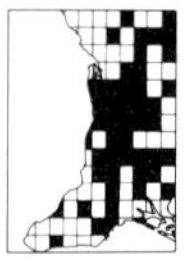

3. **Asphodelus fistulosus*, **onion weed**: 20–60 cm tall; fls (3a); 'petals' 8–12 mm long; fruit (3b) 4–7 mm long. FR, EA, EP, NL, MU, YP, KI, SE; all States; New Zealand; native to the Mediterranean and Asia.

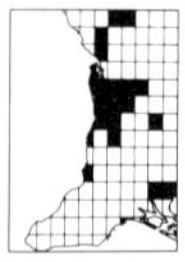

4. *Bulbine bulbosa*, **native leek**: 20–50 cm tall; fls (4a); 'petals' 10–15 mm long; fruits c. 6 mm long. FR, EP, NL, MU, YP, SE; all States except WA and NT *B. semibarbata*, **leek lily**, has 'petals' 4–6 mm long and fruits 2–5 mm long.

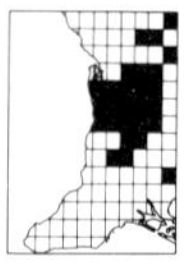

5. *Caesia calliantha*, **blue grass-lily**: 15–70 cm tall; fls (5a); 'petals' 7–10 mm long, twisting together after flowering. FR, EP, NL, MU, YP, KI, SE; all States except WA and NT. (formerly C. vittata)

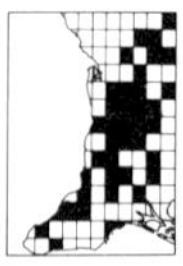

6. *Chamaescilla corymbosa*, **blue squill**: 10–15 cm tall; fls (6a); 'petals' 8–10 mm long, twisting together after flowering; fruit (6b) c. 8 mm long. FR, EP, NL, MU, YP, KI, SE; all States except NT and Qld.

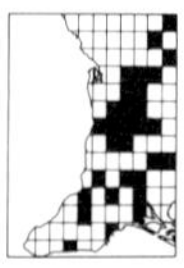

7. *Dianella revoluta*, **black-anther flax-lily**: To 1 m or more tall; fls (7a and 7b); 'petals' 6–10 mm long; fruit (7c) fleshy. NW, NU, FR, EP, NL, MU, YP, KI, SE; all States except NT. *D. longifolia*, **pale flax-lily**, has light-yellow anthers.

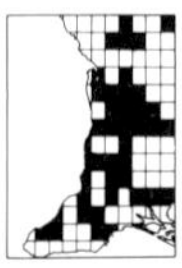

8. **Myrsiphyllum asparagoides*, **bridal creeper**: Climbing to 1–2 m; 'lvs' 2–3 cm long; fls (8a); 'petals' 5–6 mm long; fruit fleshy. FR, EP, NL, MU, YP, KI, SE; WA, NSW, Vic.; native to S. Africa. **M. declinatum* has groups of 2 or 3 very narrow 'lvs'.

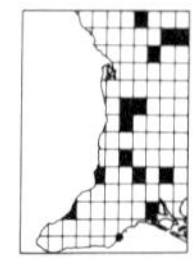

9. **Ornithogalum thyrsoides*, **chinckerinchee**: 30–70 cm tall; fls (9a); 'petals' 1.5–2.0 cm long; 3 of the anther filaments have square bases; note the true bulb. EP, MU; Vic.; native to S. Africa. **O. umbellatum*, a **star of Bethlehem**, has smaller fls and **O. arabicum*, another **star of Bethlehem**, has narrower anther filaments.

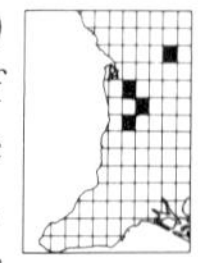

10. *Thysanotus juncifolius*, **rush fringe-lily**: 25–60 cm tall; lvs usually 0 at flowering; fls (10a); 'petals' 10–14 mm long. EP, MU, KI, SE; Qld, NSW, Vic. *T. baueri*, **mallee fringe-lily**, has lvs at flowering and petals usually 8–10 mm long.

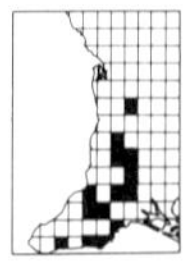

11. *Thysanotus patersonii*, **twining fringe-lily**: climbing to 10–20 cm; fls (11a); 'petals' 6–11 mm long. FR, EP, NL, MU, YP, KI, SE; all States except NT.

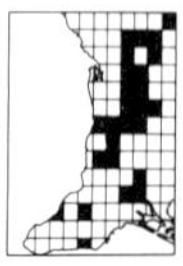

12. *Wurmbea dioica*, an **early Nancy**: 3.5–20.0 cm tall; fls (male 12a, female 12b) on separate plants; 'petals' 3.5–11.5 mm long, with a single reddish gland on each. LE, NU, GT, FR, EA, EP, NL, MU, YP, KI, SE; all States except WA and NT. *W. uniflora* has 2 glands on each 'petal' and usually only 1 fl.

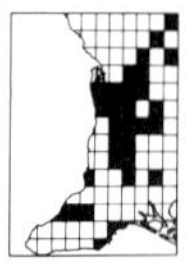

(*Continued*)

1a
1
2a
2
3a
3b
3
4a
5a
5
6
6a
6b
7c
7
7b
8a
8
9a
9
10a
10
11a
11
12a
♂
12
12 b
♀

Plate LXXIII

LILIACEAE

(Continued)

1. *Laxmannia orientalis*, **dwarf wire-lily**: 3–8 cm tall; lvs 1–3 cm long; fruit (1a) enclosed by the 6 (4 shown here) dry 'petals' which are c. 4 mm long. EP, MU, YP, KI, SE; all States except NT and Qld.

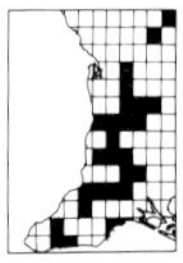

2. *Lomandra densiflora*, **pointed mat-rush**: Male (2a) and female (2b) plants 20–60 cm tall; lvs with toothed tips (2c); male fls (2d) with 'petals' c. 2.5 mm long; female fls (2e) with 'petals' c. 4 mm long. FR, EP, NL, MU, YP.

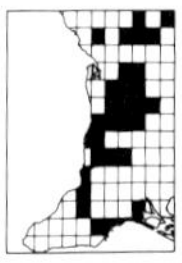

3. *Lomandra multiflora*, **stiff mat-rush**: Male (3a) and female (3b) plants 30–50 cm tall; male (3c) and female (3d) fls c. 3 mm long. LE, FR, EA, NL, MU, YP, SE; all States except WA and Tas. New Guinea.

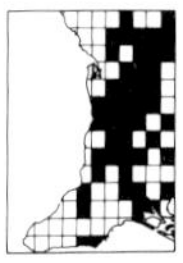

4. *Lomandra effusa*, **scented mat-rush**: Male and female plants similar c. 10–80 cm tall; lvs (4a) with notched tips; fls (male 4b, female 4c) with 'petals' 3–6 mm long. NU, FR, EP, NL, MU, YP, SE; WA, NSW, Vic.

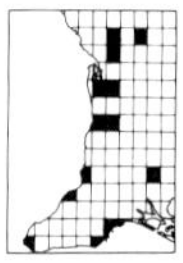

5. *Lomandra fibrata*, **Mount Lofty mat-rush**: Male (5a) and female (5b) plants 20–50 cm tall; lvs very fine and numerous, with finely toothed margins (5c); fls (male 5d, female 5e) c. 3 mm long. NL.

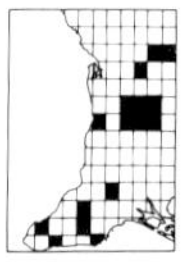

6. *Lomandra collina*, an **irongrass**: Male (6a) and female (6b) plants 5–50 cm tall; lvs (6c) with long-pointed tips; male fls 2–4 mm long; female fls 5–6 mm long. NU, EP, NL, MU, YP, SE; WA, NSW, Vic.

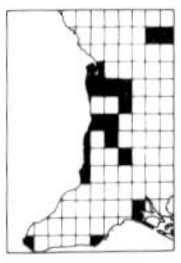

7. *Lomandra juncea*, **desert mat-rush**: Male and female plants similar, 20–50 cm tall; lvs (7a) tapering to the point; fls (male 7b, female 7c) c. 5 mm long; note the fringed bracts. EP, NL, MU, SE; Vic.

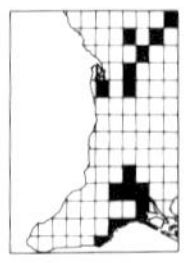

8. *Lomandra leucocephala*, **woolly mat-rush**: Male and female plants similar; infl. continuous or interrupted; lvs with pointed tips (8a); fls c. 8 mm long. NW, LE, NU, GT, EA, EP, NL, MU, YP, SE; WA, NSW, Vic.

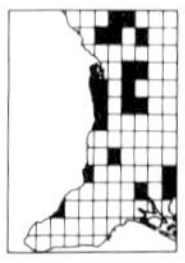

9. *Lomandra micrantha*, **small-flower mat-rush**: 20–80 cm tall; lvs (9a) pointed; male infl. (9b) more branched than the female infl. (9c); male fls (9d) 'petals' 2.0–2.5 mm long; female fls (9e) 'petals' 3–4 mm long. EP, NL, MU, YP, SE; WA, NSW, Vic.

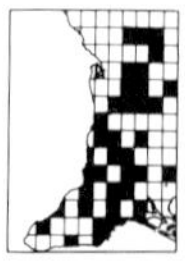

10. *Lomandra sororia*, **small mat-rush**: 4–40 cm tall; fls (male 10a, female 10b) c. 2 mm long. NL, MU, SE; Vic. 2 other species.

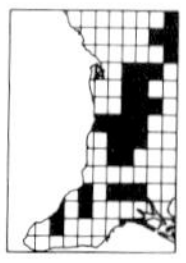

11. *Xanthorrhoea quadrangulata*, **Mount Lofty grass-tree**: Trunk to 2 m tall; lvs (11a) c. 2 mm wide; spike 5–60 cm long; fls (11b) c. 10 mm long. LE, FR, EA, NL, MU, SE.

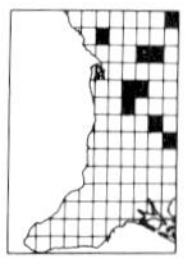

12. *Xanthorrhoea semiplana*, a **grass-tree**: Trunk usually less than 1 m tall; lvs (12a) 5–12 mm wide; spikes 65–250 cm long; fls (12b) c. 10 mm long. EA, MU, YP, KI, SE; ?Vic.

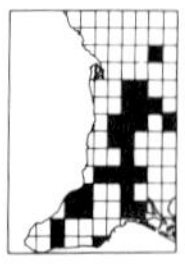

(Continued)

1a
1
2a
2d
2e
2c
2
2b
3c
3
3b
3d
3a
4c
4b
4
6c
6b
6
6a
5e
5
5d
5b
5a
5c
9b
9d
9e
9c
9a
9
7a
7b
7c
7
8a
8
10b
10a
10
11a
11
11b
12b
12a
12

LILIACEAE
(Continued)

1. *_Allium triquetrum_, **three-cornered garlic**: 2–40 cm tall; stems (1a) 3-angled; fls (1b) 12–14 mm long; onion-like smell when plant is crushed. EP, KI, SE; NSW, Vic., New Zealand; native to the Mediterranean. 6 other species.

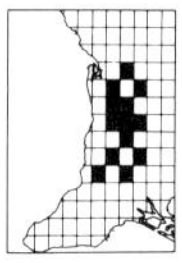

2. *_Nothoscordum gracile_, **fragrant onion**: 15–50 cm tall; stems (2a) cylindrical; fls (2b) 10–15 mm long. Differs from the *Allium* species in lacking the onion-smell and in having more (4–12) seeds in each of the 3 fruit cells. MU, SE; all States except NT; native to America.

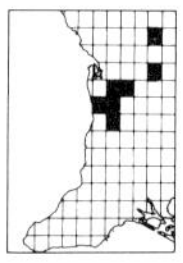

3. *Burchardia umbellata*, **milkmaids**: 15–50 cm tall; stems cylindrical; fls (3a); 'petals' 9–13 mm long; note the bract on the stems and the 3 or more bracts at the base of the infl. EP, MU, YP, KI, SE; all States except NT.

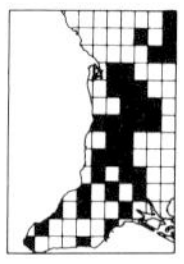

AMARYLLIDACEAE

4. *Calostemma purpureum*, **purple bells**: 14–50 cm tall; stem cylindrical; fls (4a) purplish to white, 15–18 mm long; note the green inferior ovary below the 'petals'. NW, FR, NL, MU, YP; NSW, Vic.

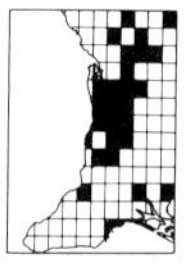

5. *_Narcissus tazetta_, **narcissus**: Usually 30–50 cm tall; fls (5a) white or yellow with a usually deeper central ring (the corona). ?EP, YP, KI; WA; native to the Mediterranean.

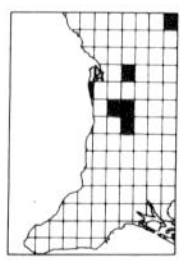

HYPOXIDACEAE

6. *Hypoxis vaginata*, **yellow star**: 5–20 cm tall; fls (6a and 6b); 'petals' 6–18 mm long; note the single 1.5–4 cm long bract on the stem. EP, SE; all States except NT and Qld.

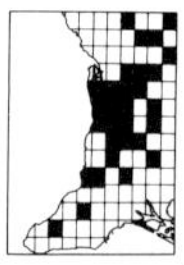

7. *Hypoxis glabella*, **tiny star**: 2–10 cm tall; fls (7a and 7b); 'petals' 4–8 mm long; note the two 4–6 mm long bracts on the stem. FR, EP, NL, MU, YP, KI, SE; NSW, Vic., Tas.; New Zealand.

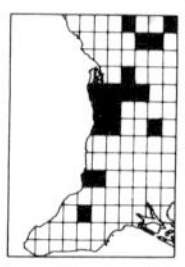

ARACEAE

8. *_Zantedeschia aethiopica_, **arum lily**: To 1 m or more tall; infl (8a) a yellow spike of very many fls surrounded by the large white spathe. Restricted to this region; WA, Vic.; native to S. Africa.

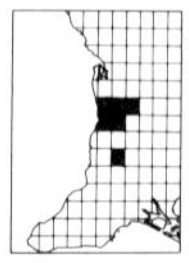

IRIDACEAE

9. *_Chasmanthe floribunda_, **Aunt Eliza**: 1.0–1.5 m tall; fls (9a) orange-red to yellow, 4.2–7.3 cm long, note the reflexed lower segments. EP, NL, YP, SE; WA, NSW, Vic.; native to S. Africa.

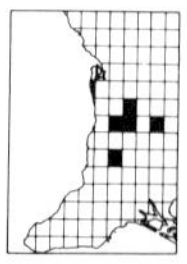

10. *_Homeria flaccida_, **one-leaved Cape tulip**: 20–50 cm tall; fls (10a) pale-orange to yellow; 'petals' 2.0–3.5 cm long; note the single lf. MU, YP, KI, SE; WA, NSW, Vic., Tas.; native to S. Africa.

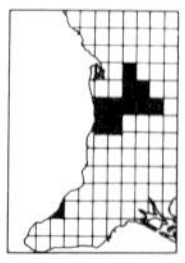

11. *_Homeria miniata_, **two-leaved Cape tulip**: 30–60 cm tall; fls (11a) always pinkish; 'petals' 1.3–2.3 cm long; note the 2 lvs. EP, NL, YP; WA, NSW, Vic.; native to S. Africa. There is a third rare species.

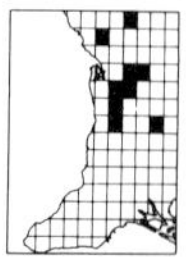

12. *_Synnotia villosa:_ 10–35 cm tall; fls (12a) 3–4 cm long incl. tube. SE; native to S. Africa.

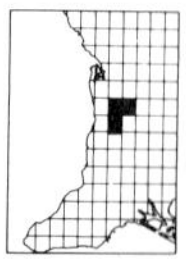

13. *_Sisyrinchium iridifolium_, **blue pigroot**: 10–60 cm tall; fls (13a); 'petals' 0.5–1.0 cm long. SE; Qld, NSW, Vic., Tas.; native to S. and central America.

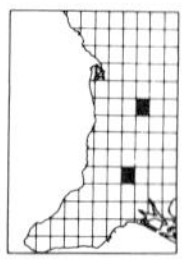

14. *_Tritonia squalida_, a **tritonia**: 30–65 cm tall; fls (14a) 3.4–4.0 cm long incl. tube. WA, Vic.; native to S. Africa.

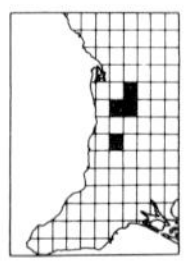

(Continued)

1a
1b
1
2b
2a
2
3a
3
4a
4
5
5a
6a
6b
6
7a
7b
7
8a
8
9a
9
10a
10
11a
11
12a
12
13a
13
14a
14

Plate LXXV

IRIDACEAE
(Continued)

1. **Babiana stricta*, **babiana**: 12–40 cm tall; lvs strongly ribbed; fls (1a) sometimes marked with red; 'petals' 3.3–4.0 cm long incl. tube. WA, NSW, Vic.; native to S. Africa.

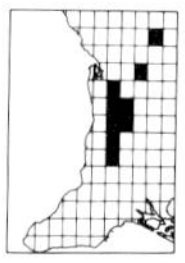

2. **Gladiolus undulatus*, **wild gladiolus**: 40–80 cm tall; fls (2a) cream to white, 'petals' 9–13 cm long incl. tube; scented at night. EP, NL, MU, KI; WA, NSW, Vic., Tas.; native to S. Africa. 5 other introduced species.

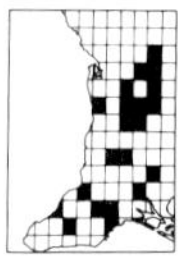

3. **Gynandriris setifolia*, **thread iris**: 4–20 cm tall; lvs to 60 cm long; fls (3a) often marked with violet; 'petals' 1.3–1.6 cm long incl. tube. Common weed. FR, EP, NL, MU, YP, KI, SE; WA, NSW, Vic.; native to S. Africa.

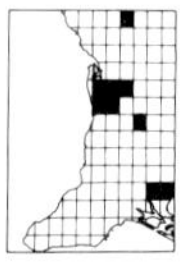

4. **Ixia polystachya*, **variable ixia**: 30–100 cm tall; fls (4a) white, pale blue, mauve or pink; 'petals' 16–39 mm incl. tube. KI; WA, NSW, Vic., Tas.; native to S. Africa. 3 other introduced species.

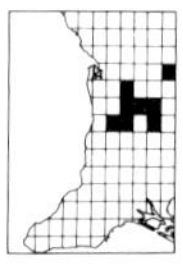

5. *Patersonia fragilis*, **swamp iris**: 30–50 cm tall; lvs convex on both faces or cylindrical, grooved; fls (5a); 'petals' 3.7–5.8 cm incl. tube. Wet heaths close to the coast. KI, SE; Qld, NSW, Vic., Tas.

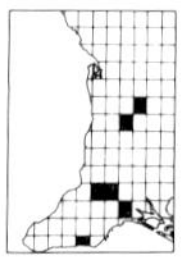

6. *Patersonia occidentalis*, **long purple-flag**: To 70 cm tall; lvs flat; fls (6a); 'petals' 4.5–7.0 cm incl. tube. KI, SE; WA, Vic., Tas.

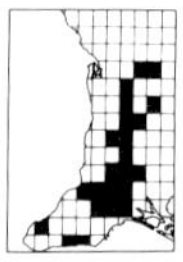

7. **Romulea minutiflora*, **lesser Guildford grass**: 2.5–6.5 cm tall; lvs long and tough; fls (7a) yellow on the tube, pale pink to violet on the lobes; 'petals' 7–12 mm long incl. tube. EP, NL, MU, YP, SE; NSW, Vic.; native to S. Africa.

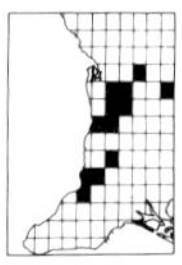

8. **Romulea rosea*, **Guildford grass**: 4.5–12.5 cm tall; lvs long and tough; fls (8a) yellow on the tube, pale to bright pink or white on the lobes; 'petals' 12–19 mm long incl. tube. EP, MU, KI, SE; WA, NSW, Vic., Tas.; native to S. Africa.

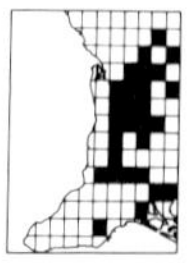

9. **Sparaxis bulbifera*, **harlequin flower**: To 55 cm tall; lvs similar to *S. tricolor*; fls pale-yellow on the tube, white to cream on the inside of the lobes, white to purplish outside; 'petals' 4.0–4.3 cm long incl. tube. Its bulbils help it to spread in the hills. EP, KI, SE; WA, NSW, Vic., Tas.; native to S. Africa.

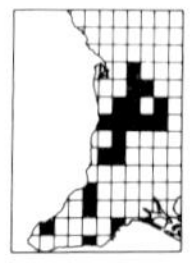

10. **Sparaxis tricolor*, **tricolor harlequin flower**: To 30 cm tall; fls (10a) bright yellow on the tube, variable in colour on the lobes; 'petals' 3.3–3.8 cm long incl. tube. NL; WA, Vic.; native to S. Africa.

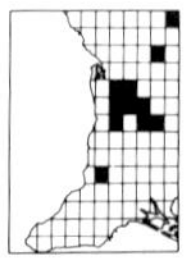

11. **Moraea bellendenii:* 50–90 cm tall; fls (11a); 'petals' 2.4–3.0 cm long; note the 3 small inner 'petals'. Native to S. Africa.

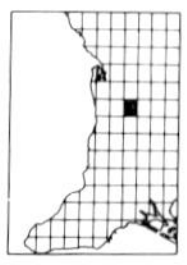

12. **Freesia* hybrid, **freesia**: 10–40 cm tall; lvs forming a well-defined fan; fls (12a); 'petals' 3–5 cm long incl. tube, scented. EP, YP, KI, SE; NSW, Vic.; all species are native to S. Africa.

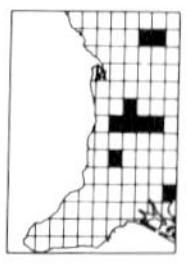

13. **Watsonia bulbillifera*, **bulbil watsonia**: 1–2 m tall; fls (13a) brick-red to salmon-pink; 'petals' 5.3–7.0 cm long incl. tube; producing numerous bulbils on the lower part of the infl. NL, KI; all States except NT; native to S. Africa.

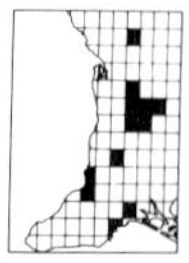

1
1a
2
2a
3
3a
4
5
5a
6
6a
7
7a
8
8a
9
9
10
10
10a
11
11a
12
12a
13
13a

Plate LXXVI

JUNCACEAE

1. **Juncus acutus*, **sharp rush**: To 130 cm tall; stem pith (1a) continuous; lvs stem-like; fls 2.5–4.0 mm long; fruit (1b) longer than the fls. FR, EP, NL, MU, SE; WA, NSW, Vic.; New Zealand; native to Europe, Africa and the Americas.

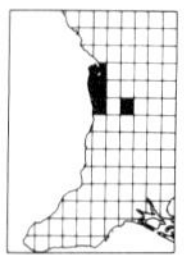

2. **Juncus articulatus*, **jointed rush**: 15–50 cm tall; lvs compressed, hollow with cross-walls (2a); fls (2b) c. 3 mm long. NL, MU, SE; WA, NSW, Vic., Tas.; New Zealand; native to Europe, Asia and the Americas.

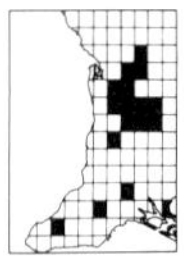

3. *Juncus australis*, **austral rush**: 60–120 cm tall; stem pith (3a) interrupted by air spaces; lvs 0; fls (3b) 2.2–3.0 mm long. SE; NSW, Vic., Tas.; New Zealand.

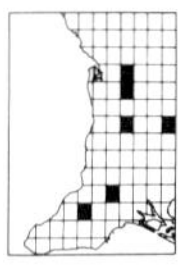

4. *Juncus holoschoenus*, **joint-leaved rush**: Usually less than 45 cm tall; stem hollow with cross-walls; lvs (4a) stem-like, with cross-walls; fls (4b) 3–4 mm long. FR, SE; all States except NT and Qld.; New Zealand.

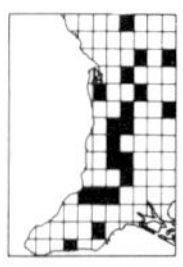

5. *Juncus kraussii*, **sea rush**: About 1 m tall; stem pith (5a) continuous; lvs stem-like; fls (5b) c. 3 mm long; fruit shorter than the fls. LE, FR, EP, NL, MU, YP, KI, SE; all States; New Zealand.

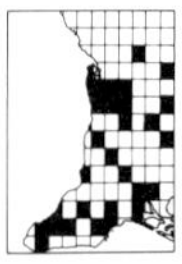

6. *Juncus pallidus*, **pale rush**: 50–200 cm tall; stem pith (6a) continuous; lvs 0; fls (6b) 2.6–3.5 mm long. NL, KI, SE; all States except NT and Qld; New Zealand.

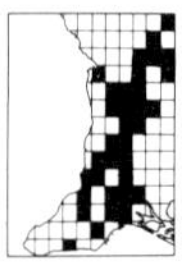

7. *Juncus pauciflorus*, **loose-flower rush**: 25–100 cm tall; stem pith (7a) usually continuous; fls 2–3 mm long; fruit (7b) longer than the fls. MU, KI, SE; all States except NT and Qld; New Zealand.

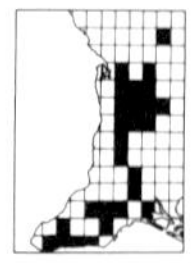

8. *Juncus planifolius*, **broad-leaved rush**: 10–30 cm tall; stem pith (8a) continuous; lvs flattish; fls (8b) 1.5–2.5 mm long. FR, KI, SE; all States except NT; New Zealand, S. America.

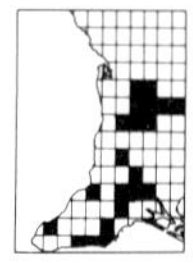

9. *Juncus prismatocarpus*, **branching rush**: Usually 30–60 cm tall; stem pith interrupted; lvs (9a) compressed, hollow with cross-walls; fls (9b) 3–4 mm long. FR, MU, SE; all States except NT; New Zealand.

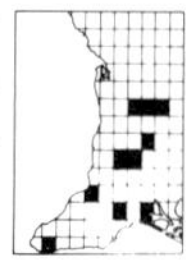

10. *Juncus sarophorus*, a **rush**: 70–120 cm tall; stem pith (10a) interrupted; lvs 0; fls 2.0–2.5 mm long. FR, NL, MU; NSW, Vic., Tas.; New Zealand.

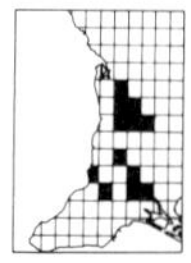

11. *Juncus subsecundus*, **finger rush**: 20–90 cm tall; stem pith (11a) interrupted; lvs 0; fls (11b) 2–4 mm long. LE, FR, EP, NL, MU, SE; WA, Qld, NSW, Vic., Tas.

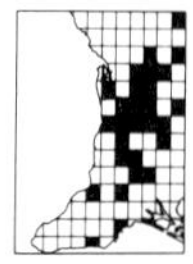

12. *Juncus usitatus*, **common rush**: Usually 35–120 cm tall; stem pith (12a) interrupted; lvs 0; fls (12b) 1.3–2.0 mm long. EA, MU; WA, Qld, NSW, Vic.; New Zealand, New Caledonia.

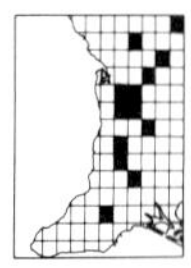

(Continued)

2b
3b
4b
5b
6b
3a
1a
2a
4a
5a
6a
1
2
3
4
5
6
7b
7a
8a
9a
10a
11a
12a
7
8
8b
9b
9
10b
10
11b
11
12b
12

Plate LXXVII

JUNCACEAE
(Continued)

1. *Juncus bufonius*, **toad rush**: Annual 2–30 cm tall; stem pith (1a) continuous; lvs flat at base and on stems; fls (1b) 6–7 mm long. NW, GT, FR, EP, NL, MU, YP, KI, SE; all mainland States; cosmopolitan.

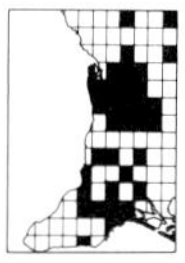

2. *Juncus caespiticius*, **grassy rush**: Perennial 10–40 cm tall; stem pith (2a) continuous; lvs flat, only at base of stems; fls (2b) 2.5–3.0 mm long. FR, EP, NL, KI, SE; all States except Qld and NT; New Zealand.

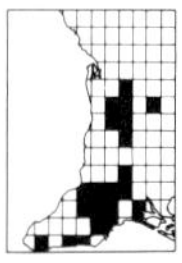

3. **Juncus capitatus*, **capitate rush**: Annual 2–8 cm tall; stem pith (3a) continuous; lvs flat, only at the base of the stems; fls (3b) 4–6 mm long. EP, NL, MU, YP, KI, SE; WA, NSW, Vic., Tas.; native to Europe, Asia, Africa and S.America. 5 other species.

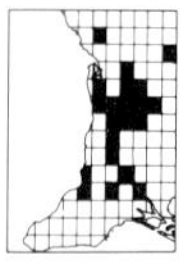

Note: Many of the 20 species of *Juncus* found in the area covered in this book are extremely difficult to identify.

4. *Luzula meridionalis*, a **woodrush**: 20–50 cm tall; fls (4a) c. 3 mm long; note hairy leaves. FR, EP, NL, MU, KI, SE; WA, NSW, Vic., Tas. 3 other species.

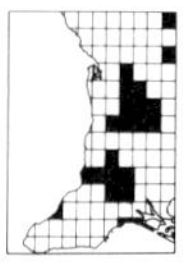

RESTIONACEAE

5. *Empodisma minus:* 25–200 cm tall; male plants (5a) with sessile spikelets; female plants (5b) with flowers with 2- or 3-branched styles. KI, SE; Qld, NSW, Vic., Tas.; New Zealand.

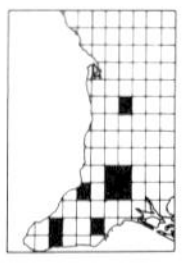

6. *Hypolaena fastigiata*, **tassel rope-rush**: 20–40 cm tall; male plants (6a) with stalked spikelets; female plants (6b) with flowers with 2- or 3-branched styles. FR, EP, MU, KI, SE; all States except NT.

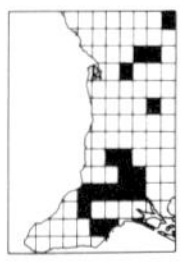

7. *Lepidobolus drapetocoleus*, **scale shedder**: c. 30 cm tall; male (7a) and female (7b) plants with terminal heads; scales along stems soon falling off. YP, SE; Vic.

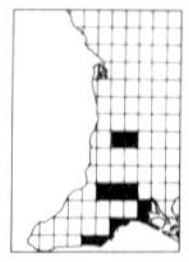

8. *Leptocarpus brownii*, **coarse twine-rush**: 50–100 cm tall; spikelets on male plants (8a) 5–12 mm long; spikelets on female plants (8b) in a compact infl. EP, SE; NSW, Vic., Tas.

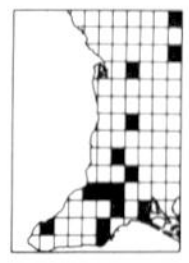

9. *Leptocarpus tenax*, **slender twine-rush**: 50–100 cm tall; spikelets on male plants (9a) 2–4 mm long; spikelets on female plants (9b) in a compact infl. MU, KI, SE; all States except NT. The only other species of Restionaceae is the very rare *Lepyrodia valliculae* which has unbranched stems and stalkless fls in a branched infl.

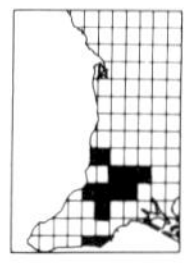

XYRIDACEAE

10. *Xyris operculata*, **tall yellow-eye**: 30–50 cm tall; fls (10a) to 20 mm across. KI, SE; all States except WA and NT.

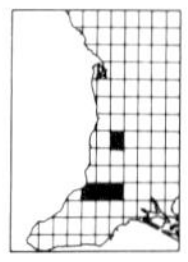

CENTROLEPIDACEAE

11. *Aphelia gracilis*, **slender aphelia**: 2–4 cm tall; spikelets (11a) 3–4 mm long, generally to one side of the line of its stem. KI, SE; NSW, Vic., Tas.

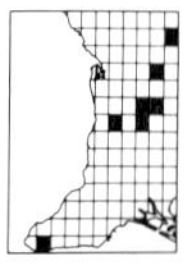

12. *Aphelia pumilio*, **dwarf aphelia**: 1–3 cm tall; spikelets (12a) 4–6 mm long, continuing the line of its stem. KI, SE, Vic., Tas.

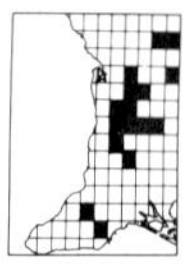

(Continued)

1b
2b
3b
1a
2a
3a
4
4a
5a
5b
6b
6a
6
5
1
2
3
7b
7a
8b
8a
9a
9b
10a
11a
12a
7
8
9
10
11
12

Plate LXXVIII

CENTROLEPIDACEAE
(Continued)

1. *Centrolepis aristata*, **pointed centrolepis**: 2–10 cm tall; note absence of hairs and long points to both bracts of the head. EP, NL, MU, YP, KI, SE; NSW, Vic., Tas.

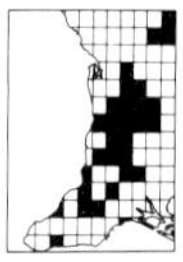

2. *Centrolepis fascicularis*, **tufted centrolepis**: 2.5–6.0 cm tall; note presence of hairs; perennial. KI, SE; all States except NT; New Guinea.

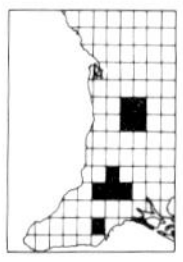

3. *Centrolepis polygyna*, **wiry centrolepis**: 1.5–6.0 cm tall; note absence of hairs and long points on only one of the bracts of the head. FR, EP, NL, MU, YP, KI, SE; WA, NSW, Vic., Tas.

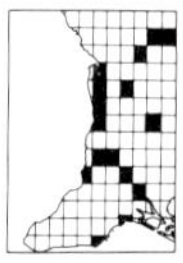

4. *Centrolepis strigosa*, **hairy centrolepis**: 2–7 cm tall; note presence of hairs, annual. FR, EP, NL, MU, YP, KI, SE; all States except NT. *C. cephaloformis*, **dwarf centrolepis**, bears its heads at the base of the plant.

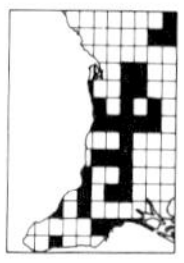

ALISMATACEAE

5. **Alisma lanceolatum*, **water plantain**: To 1 m tall; fls (5a) to 9 mm diam. Grows in damp ground or shallow water. MU; Vic.; native to the Northern Hemisphere.

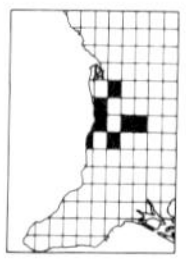

APONOGETONACEAE

6. **Aponogeton distachyos*, **Cape pond lily**: Lf blades 6–20 cm long, floating in fresh water; each of 2 opposite spikes to 6 cm long; each fl (6a) with 1 petal-like structure which is white or pink and c. 15 mm long, scented; fruit (6b). Restricted to this region. NSW, Vic.; native to S. Africa.

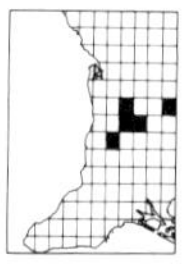

JUNCAGINACEAE

7. *Triglochin centrocarpum*, **dwarf arrowgrass**: 5–15 cm tall; fruit (7a) 3–5 mm long, note the spurs pointing downwards from the fruit base. All regions of SA; all States. *T. calcitrapum*, **spurred arrowgrass**, has its spurs curving upwards. 2 other small species differ on fruit characters.

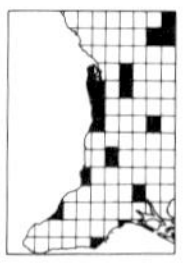

8. *Triglochin procerum*, **water ribbons**: 10–50 cm or more tall; fruit (8a) 1–2 cm long; lvs to 2 m long, erect or floating in fresh water. LE, MU, KI, SE; WA, NSW, Vic.

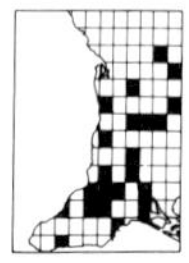

9. *Triglochin striatum*, **streaked arrowgrass**: 10–30 cm tall; fruit (9a) c. 2 mm long; lvs 2–30 cm long, erect; in wet places. LE, FR, EP, NL, MU, YP, KI, SE; all States; New Zealand, the Americas, Africa.

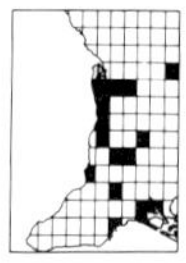

POTAMOGETONACEAE

10. *Potamogeton pectinatus*, **fennel pondweed**: Stems to 3 m long in fresh water; lvs narrow, 1.5–15.0 cm long; fruit (10a) 2.4–4.0 mm long. LE, EP, NL, MU, KI, SE; all States except NT; cosmopolitan. 4 other species, some with broad lf blades.

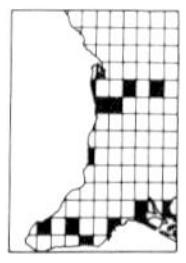

11. *Ruppia megacarpa*, **widgeon grass**: Stems to 2.5 m long in salty water; lvs narrow, 5–25 cm long; lf tip variations (11a); fruit (11b) 3–4 mm long. FR, EP, NL, MU, KI, SE; WA, NSW, Vic.; New Zealand. 2 other species.

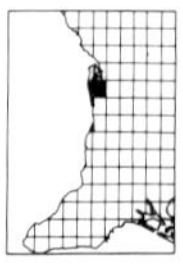

CYMODOCEACEAE

12. *Amphibolis antarctica*, a **wire weed**: Stems to 1.5 m long in the sea along the coast; lvs 18–50 mm long; male (12a) and female (12b) fls inconspicuous; young seedlings (12c) with a 'grappling apparatus' at their base for anchoring. NU, EP, YP, KI, SE; WA, Vic., Tas. Common on beaches. *A. griffithii* has narrower leaves at least 10 times as long as broad. 4 *Posidonia* species, tape weeds, POSIDONIACEAE, with long tape-like lvs, are also washed up on the beaches.

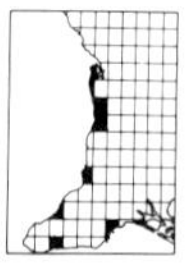

1
2
3
4
5
5a
6
6a
6b
7
7a
8
8a
9
9a
10
10a
11
11a
11b
12
12a
12b
12c
♂
♀

Plate LXXIX

GRAMINEAE

1. *Cymbopogon ambiguus*, **scented grass**: 30–200 cm tall; infl 20–25 cm long; spikelets (1a) 1.5–2.0 cm long incl. awn. Pleasant scent. NW, LE, GT, FR, EA, EP, NL, MU; all States except Vic and Tas. *C. obtectus*, silky heads, has broader leaves and the inflorescences reflex with age.

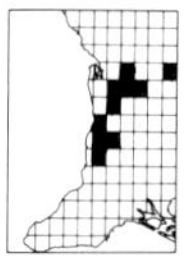

2. *Dichanthium sericeum*, **silky blue-grass**: 30–80 cm tall; infl branches 2.5–7.0 cm long; spikelets (2a) 22–26 mm long incl. awn. NW, LE, GT, FR, EA, NL; all mainland States.

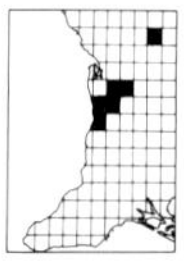

3. *Chloris truncata*, **windmill grass**: 16–45 cm tall; infl branches usually 9–13 cm long; spikelets (3a) 13–19 mm long incl. awn, black when mature. LE, FR, EA, EP, NL, MU, YP, SE; all States except NT and Tas. 3 other spp.

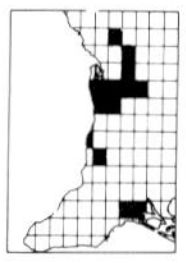

4. **Cynodon dactylon*, **couch-grass**: Common lawn grass, to 30 cm tall if not cut; infl branches 2–5 cm long; spikelets (4a) c. 2 mm long. NW, LE, EA, FR, NL, MU, YP, SE; cosmopolitan and widespread in Australia. 1 other rare species.

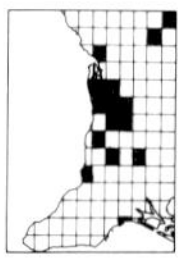

5. **Digitaria ciliaris*, a **crab grass**: Stems creeping, to 70 cm tall; infl branches 5–16 cm long; spikelets (5a) c. 3 mm long, greenish at first. Garden weed. EP, MU, SE; all States; native to the tropics. 6 other species.

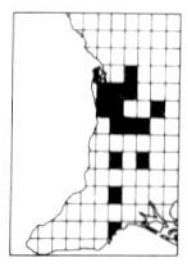

6. *Spinifex sericeus*, **rolling spinifex**: 30–60 cm tall; male infl. (6a) branches 5–8 cm long; male spikelets (6b) 8–12 mm long; bisexual infl. (6c) branches 10–25 cm long; bisexual spikelets (6d) c. 2 cm long. Common on coastal dunes. NU, EP, YP, KI, SE; all States except WA and NT.

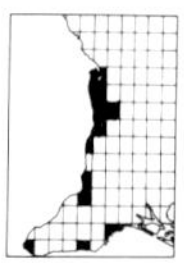

7. **Paspalum dilatatum*, **paspalum**: 50–100 cm tall; infl branches 4–10 cm long; spikelets (7a) 3–4 mm long. Garden weed. FR, EP, NL, MU, YP, KI, SE; all States; native to S. America.

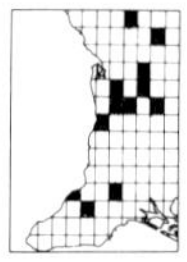

8. **Paspalum distichum*, **water couch**: Usually less than 45 cm tall; infl branches usually 4–6 cm long; spikelets (8a) c. 3 mm long. Mainly in wet places. EP, NL, MU, YP, SE; all States; native to warm parts of the world. *P. vaginatum*, **salt-water couch**, occurs in coastal salty areas.

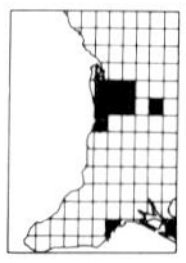

9. *Panicum effusum*, **hairy panic**: 20–50 cm tall; infl 10–20 cm long; spikelets (9a) 2.0–2.5 mm long. NW, LE, NU, GT, FR, EA, EP, NL, MU, YP, SE; all mainland States. 4 other species.

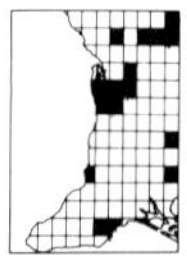

10. **Piptatherum miliaceum*, **rice millet**: To 1 m or more tall; infl 30–40 cm long; spikelets (10a) 4–5 mm long incl. awn. Large weedy grass in large clumps. FR, EP, NL, MU, YP, SE; all States except Qld; native to the Mediterranean.

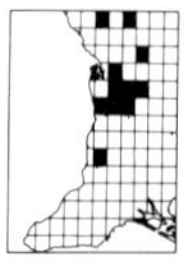

11. **Sorghum halepense*, **Johnson grass**: 50–150 cm tall; infl usually 17–23 cm long; spikelets (11a) c. 5 mm long incl. awn which is not always present. LE, EA, NL, MU, YP, KI, SE; all mainland States; native to the Mediterranean. 2 other species have occurred as escapes.

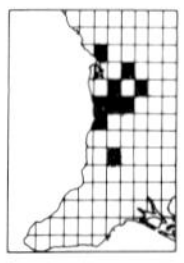

12. **Echinochloa crus-galli*, **barnyard-grass**: 30–80 cm tall; infl 6–20 cm long; spikelets (12a) 3–4 mm long plus awn 0–30 mm long. LE, EA, EP, NL, MU, YP, SE; all States except NT; almost cosmopolitan. 2 other species.

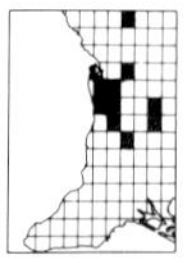

(*Continued*)

1a
1
2
2a
3
3a
4a
4
5
5a
6a
6b
6
6c
6d
7a
7
8
8a
9a
9
10a
10
11a
11
12
12a

Plate LXXX

GRAMINEAE

(Continued)

1. *Agrostis aemula*, a **blown grass**: 25-60 cm tall; infl 15-30 cm long; spikelets (1a) 4-5 mm long excl. awn; lemma hairy. ?NW, ?FR, EP, NL, MU, YP, KI, SE; all States except NT.

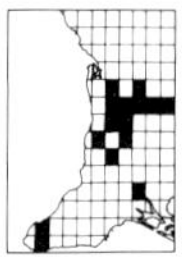

2. *Agrostis avenacea*, a **blown grass**: 20-70 cm tall; infl 5-30 cm long, lower branches drooping with age; spikelets (2a) 2.5-3.5 mm long excl. awn; lemma hairy. NW, LE, GT, FR, EA, EP, MU, YP, KI, SE; all States; New Zealand.

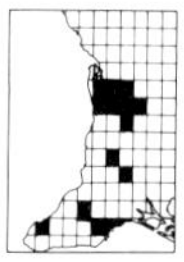

3. *Agrostis billardieri*, a **blown grass**: 15-60 cm tall; infl 10-30 cm long; spikelets (3a) 5-7 mm long excl. awn. Grows on dunes. EP, NL, MU, YP, KI, SE; Vic., Tas.; New Zealand.

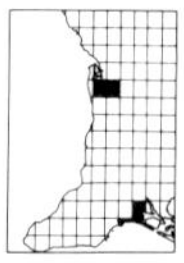

4. **Agrostis capillaris*, **brown-top bent**: Usually 20-50 cm tall; infl 1-20 cm long; spikelets (4a) 2.0-3.5 mm long excl. awn, no hairs at base of lemma, awn sometimes 0. MU, SE; all States except NT and Qld; New Zealand; native to the Northern Hemisphere. *A. gigantea*, **red-top bent**, has a longer ligule 5-6 mm.

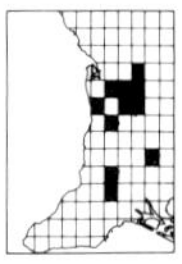

5. **Aira caryophyllea*, a **silvery hair-grass**: Usually 10-30 cm tall; infl 2-10 cm long; spikelets (5a) 2.5-3.0 mm long excl. awn; usually 2 awns. FR, EP, NL, MU, YP, KI, SE; all States except NT; native to Europe, Africa and Asia.

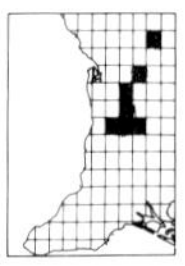

6. **Aira cupaniana*, a **silvery hair-grass**: Usually 10-30 cm tall; infl 2-10 cm long; spikelets (6a) 2.5-3.0 mm long excl. awn; often 1 awn. *A. elegantissima*, delicate hair-grass usually has 1 awn but pedicels 2-5 times as long as the spikelets. FR, EP, NL, MU, YP, KI, SE; all States except NT and Qld; native to Europe and Asia.

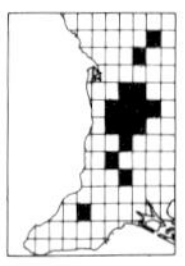

7. **Avena barbata*, **bearded oat**: 0.4-1.0 m tall; infl 10-40 cm long; spikelets (7a) 14-25 mm long excl. awns; note the long bristles on either side of the base of the awn. Common weed. LE, GT, FR, EP, NL, MU, YP, KI, SE; WA, NSW, Vic.; native to the Mediterranean.

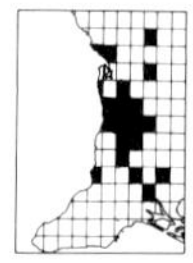

8. **Avena sativa*, **cultivated oat**: 30-60 cm tall; infl to 20 cm long; spikelets (8a) usually 25-30 mm long excl. awns (which are not always present); lemma lacking hairs. NW, LE, EP, NL, YP, KI, SE; all States except NT; native to Europe. 2 other species.

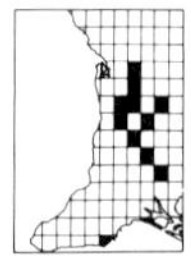

9. **Briza maxima*, **quaking-grass**: 20-60 cm tall; infl 2-6 cm long; spikelets (9a) 10-20 mm long. FR, EP, NL, MU, KI, SE; all States except NT; native to the Mediterranean.

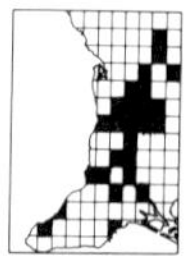

10. **Briza minor*, **shivery grass**: Usually 10-50 cm high; infl 2-13 cm long; spikelets (10a) 2-5 mm long. FR, EP, NL, YP, KI, SE; all States except NT; native to Europe.

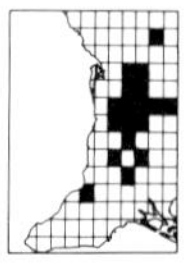

11. *Deschampsia caespitosa*, **tufted hair-grass**: 30-100 cm or more tall; infl 15-25 cm long; spikelets (11a) 5-6 mm long excl. awns. Restricted to this region; NSW, Vic., Tas.; New Zealand, Europe, Asia, Africa, N. America. Temperate and arctic regions.

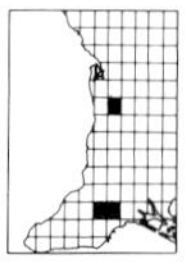

12. *Microlaena stipoides*, **weeping rice-grass**: To 60 cm tall; infl 8-20 cm long; spikelets (12a) c. 10 mm long plus awns c. 20 mm long; note the separation of the glumes from the lemmas. FR, NL, MU, KI, SE; all States; New Zealand.

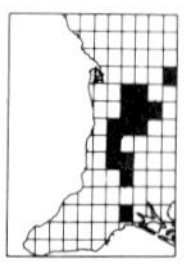

(Continued)

1
1a
2
2a
3
3a
4
4a
5
5a
6
6a
7
7a
8
8a
9
9a
10
10a
11
11a
12
12a

Plate LXXXI

GRAMINEAE

(*Continued*)

1. **Poa annua*, **winter grass**: 8–30 cm tall; infl 1–12 cm long; spikelets (1a) 3.0–4.5 mm long; annual. LE, FR, EP, NL, MU, YP, KI, SE; all States except NT; native to Europe; now cosmopolitan. *Poa fax*, scaly poa, has more (9–11) nerves on the lemmas.

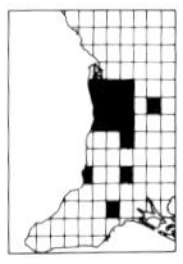

2. **Poa bulbosa*, **bulbous meadow-grass**: 5–45 cm tall; infl c. 2–6 cm long; spikelets (2a) 4–5 mm long; bulbous-based. FR, EP, NL, MU, YP, KI, SE; WA, NSW, Vic.; native to Europe and the Mediterranean.

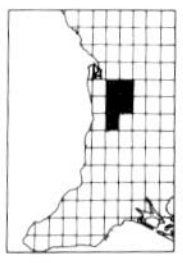

3. *Poa clelandii*, a **poa**: To 75 cm tall; infl to 20 cm long; spikelets (3a) 3.5–5.0 mm long; lower leaf sheaths usually purplish. EP, NL, MU, ?YP, SE; Vic.

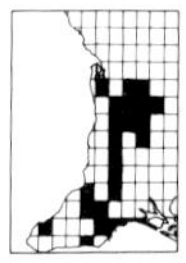

4. *Poa crassicaudex*, a **poa**: To 90 cm tall; infl to 17 cm long; spikelets (4a) 4–5 mm long; base swollen and often yellowish. FR, EP, NL, MU, YP, KI, SE; Vic. *P. drummondiana*, shaking grass, has a similarly swollen base but lacks hairs on the leaf sheaths and has broader spikelets.

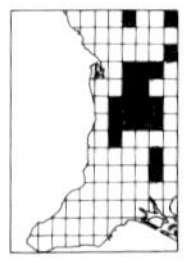

5. *Poa labillardieri*, a **tussock grass**: 30–120 cm tall; infl 10–25 cm long; spikelets (5a) 4–5 mm lóng; leaves rough but otherwise rather like *P. poiformis*; inland. FR, EP, NL, MU, YP, SE; all States except WA and NT.

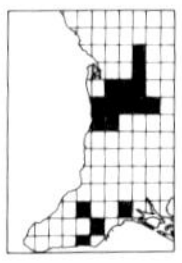

6. *Poa poiformis*, a **tussock grass**: 20–90 cm tall; infl 8–30 cm long; spikelets (6a) 4.5–6.0 mm long; leaves smooth but otherwise rather like *P. labillardieri*; coastal. ?FR, EP, NL, MU, YP, KI, SE; all States except NT and Qld; Lord Howe Island. *P. halmaturina*, **poa**, has rather rigid leaves as often occurs in *P. poiformis* but the plant is smaller and has slender rhizomes.

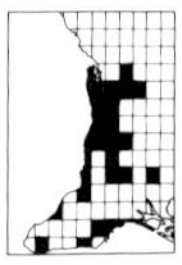

7. **Poa pratensis*, **Kentucky blue-grass**: 10–90 cm tall; infl to 30 cm long; spikelets (7a) 2.5–6.0 mm long. A lawn and fodder species. ?FR, KI, SE; all States except NT; native to Europe and Asia. Commonly planted in lawns and used as fodder grass. *P. fordeana*, **Forde poa**, also has a ligule more than 2 mm long and thinly membranous but spikelets up to 11 mm long.

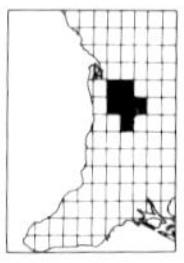

8. *Poa umbricola*, a **poa**: 30–50 cm tall; infl 2.5–9.0 cm long; spikelets (8a) c. 3.5 mm long; note the leafy stems (8b). Restricted to this region. *P. tenera*, **slender tussock grass**, also has leafy stems but the leaves tend to be finer.

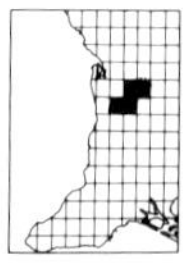

9. **Ehrharta calycina*, **perennial veldt grass**: 30–70 cm tall; infl usually less than 15 cm long; spikelets (9a) 4–8 mm long. EP, NL, MU, YP, KI, SE; WA; native to southern Africa. **E. erecta*, panic veldt grass, lacks hairs on the lemmas and has cross-bars on 1 lemma.

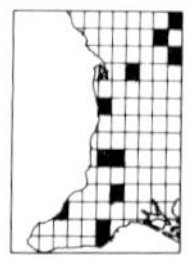

10. **Ehrharta longiflora*, **annual veldt grass**: To 90 cm tall; infl to 20 cm long; spikelets (10a) usually 10–18 mm long incl. awns. NU, FR, EP, MU, YP, KI, SE; Qld, NSW, Vic.; native to southern Africa.

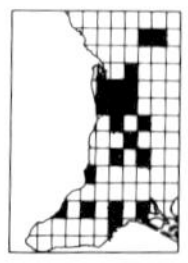

11. **Ehrharta villosa*, **pyp grass**: To 120 cm tall; infl 4–20 cm long; spikelets (11a) 12–15 mm long incl. awns. EP, NL, MU, YP; WA, Vic.; native to southern Africa. Differs from *E. longiflora* in having longer hairs on the lemmas.

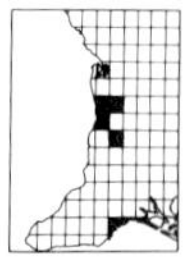

12. *Amphibromus archeri*, **pointed swamp wallaby-grass**: To 1.2 m tall; infl to 35 cm long; spikelets (12a) 12–18 mm long. SE; Vic., Tas. 2 other species lacking the long bristles on either side of the awns are hard to distinguish.

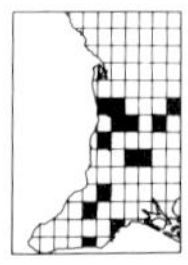

(*Continued*)

1a
1
2a
2
3a
3
4a
4
5
5a
6a
6
7a
7
8b
8
8a
9a
9
10a
10
11a
11
12a
12

Plate LXXXII

GRAMINEAE
(Continued)

1. *Stipa blackii*, **crested spear-grass**: To 1.5 m tall; infl to 50 cm long; spikelets (1a) 10–17 mm long excl. awns; note tufts of long hairs at base of awn which is shorter than in most species (3–4 cm) and twice bent. FR, EA, EP, NL, MU, SE; WA and Vic.

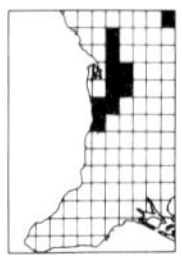

2. *Stipa curticoma*, a **spear-grass**: To 1.2 m tall; infl 15–30 cm long; spikelets (2a) 12–16 mm long excl. awns; lemma has a short tuft of hairs at the base of the awn and white hairs all over. EP, NL, MU, YP, KI, SE; Vic.

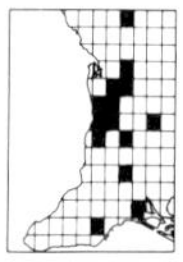

3. *Stipa drummondii*, **cottony spear-grass**: To 1 m tall; infl to 40 cm long; spikelets (3a) 7–12 mm long excl. awns; note long hairs on leaves, sickle-shaped awn with 1 mm long hairs on its lower parts. NU, FR, EA, EP, NL, MU, YP, SE; WA,

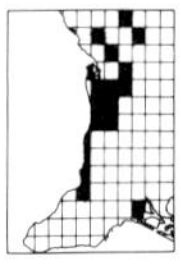

4. *Stipa elegantissima*, **elegant spear-grass**: To 2 m tall; infl 15–30 cm long; spikelets (4a) 15–25 mm long excl. awns; note long hairs on infl branches. NU, FR, EA, EP, NL, MU, YP, KI; WA, NSW, Vic.

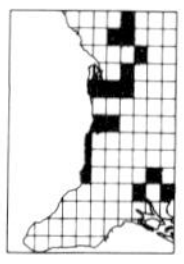

5. *Stipa eremophila*, **desert spear-grass**: To 85 cm tall; infl 15–30 cm long; spikelets (5a) 15–25 mm long excl. awns; note twice-bent awns and deep orange hairs on the lemma shorter above. ?NW, NU, GT, FR, EA, EP, NL, MU, YP, SE; WA, NSW, Vic.

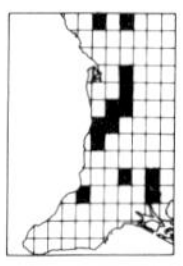

6. *Stipa flavescens*, a **spear-grass**: To 1.2 m tall; infl 10–40 cm long; spikelets (6a) 9–14 mm long excl. awns; lemma has a short tuft of hairs at the base of the awn and yellowish hairs all over. EP, NL, MU, YP, KI, SE; all States except NT and Qld.

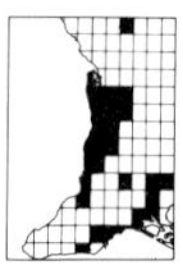

7. *Stipa mollis*, **soft spear-grass**: To 1 m tall; infl 10–30 cm long; spikelets (7a) 15–20 mm long excl. awns; awn twice-bent with hairs 1–4 mm long in a spiral along the lower part. FR, EP, NL, MU, YP, KI, SE; all States except NT and Qld.

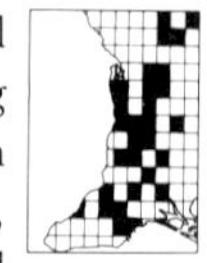

8. *Stipa muelleri*, **wiry spear-grass**: To 90 cm tall; infl 2.0–3.5 cm long; spikelets (8a) 18–30 mm long excl. awns; note only 1–3 spikelets in an infl KI, SE; Vic.

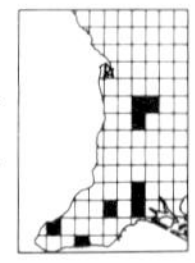

9. *Stipa nodosa*, a **spear-grass**: To 1 m tall; infl 15–50 cm long; spikelets (9a) 10–15 mm long; note sickle-shaped awn with very short hairs and leaves 1–3 mm broad. NW, LE, GT, FR, EA, EP, NL, MU, YP, KI, SE; all States except WA and NT.

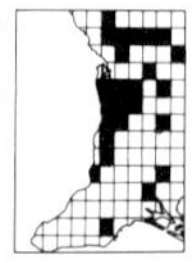

10. *Stipa scabra*, a **spear-grass**: To 50 cm tall; infl to 30 cm long; spikelets (10a) usually 9–10 mm long excl. awns; note sickle-shaped awn with very short hairs and leaves usually to 1.5 mm broad. NW, NU, FR, EA, EP, NL, MU, YP, KI, SE; all States except NT and Tas.

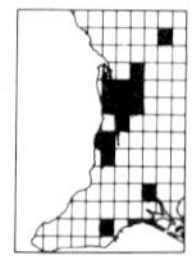

11. *Stipa semibarbata*, **fibrous spear-grass**: To 1 m tall; infl to 30 cm long; spikelets (11a) 18–25 mm long excl. awns; similar to *S. mollis* but hairs to 1 mm long and evenly distributed along lower part of awn. EA, EP, NL, MU, YP, KI, SE; NSW, Vic., Tas.

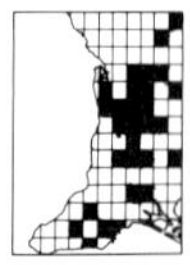

12. *Stipa setacea*, **corkscrew grass**: To 80 cm tall; infl 10–20 cm long; spikelets (12a) 9–15 mm long excl. awns; awns twice-bent, microscopically hairy; ligule long (2–9 mm). FR, EA, NL, MU, YP, SE; Qld, NSW. 19 other species.

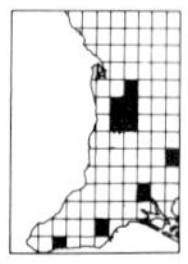

(Continued)

1
1a
2
2a
3
3a
4
4a
5
5a
6
6a
7
7a
8
8a
9
9a
10
10a
11
11a
12
12a

Plate LXXXIII

GRAMINEAE

(Continued)

1. ******Desmazeria rigida*, **fern grass**: 5–20 cm tall; infl 4–8 cm long; spikelets (1a) 5–6 mm long; lemmas rounded on their backs. FR, EP, NL, MU, YP, KI, SE; all States except NT and Qld; native to Europe and Asia.

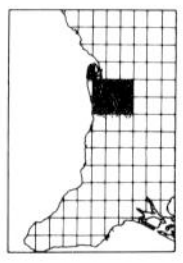

2. *Deyeuxia densa*, a **bent-grass**: 50–100 cm tall; infl 8–18 cm long; spikelets (2a) 4.5–5.0 mm long excl. awn. KI, SE; Vic. and Tas. *D. minor*, a **bent-grass**, also has the awn attached near the middle of the lemma but the spikelet is c. 4 mm long.

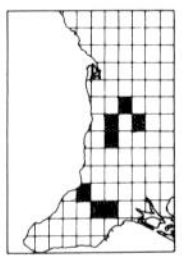

3. *Deyeuxia quadriseta*, **reed bent-grass**: To 1 m tall; infl 5–15 cm long; spikelets (3a) 4.5–5.0 mm long excl. awns; awn attached near the base of the lemma. FR, EP, YP, KI, SE; all States except NT and Qld.

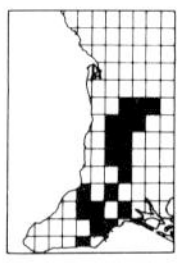

4. *Dichelachne longiseta*, **long-hair plume-grass**: To 1.2 m tall; infl to 20 cm long; spikelets (4a) 5–11 mm long. FR, EP, NL, MU, YP, KI, SE; Australian distribution not available. *D. inaequiglumis*, a **plume-grass**, has a much sparser infl

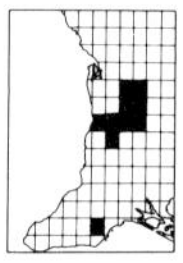

5. *Dichelachne micrantha*, **short-hair plume-grass**: To 1.1 m tall; infl to 19 cm long; spikelets (5a) 3–5 mm long. FR, NL, KI, SE; all States except NT; New Zealand.

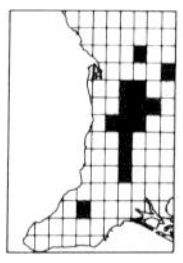

6. *Eragrostis benthamii*, **common lovegrass**: 11–60 cm tall; infl 8–30 cm long; spikelets (6a) 5–12 mm long. NL, SE; all mainland States.

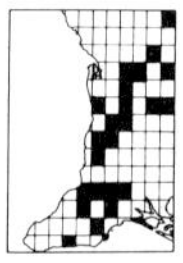

7. **Eragrostis cilianensis*, **stinkgrass**: 12–90 cm tall; infl 4–30 cm long; spikelets (7a) 6–18 mm long; note (7b) glands along lf margins. NW, LE, GT, FR, EA, EP, NL, MU, YP, SE; all States; native to Europe, Africa and Asia.

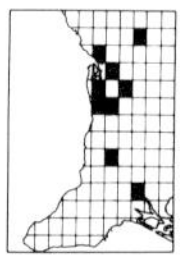

8. **Eragrostis curvula*, **African lovegrass**: 30–120 cm tall; infl 6–30 cm long; spikelets (8a) 4–10 mm long. 8 other species. EP, MU, YP, SE; all mainland States; native to southern Africa.

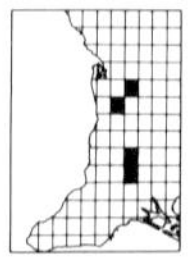

9. **Festuca arundinacea*, **tall meadow fescue**: To over 1 m tall; infl 15–30 cm long; spikelets (9a) 8–12 mm long; lemmas angled along their backs. Introduced as a pasture grass. MU, SE; all States except NT; native to Europe and Asia. 3 or 4 other species.

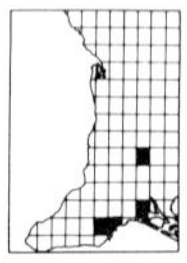

10. **Holcus lanatus*, **Yorkshire fog**: To 1 m tall; infl 8–15 cm long; spikelets (10a) 4–5 mm long; the pink mature infl often conspicuous along roadsides. MU, SE; all States except NT; native to Europe, Asia and N. Africa. 1 other rare species, *H. setiglumis*.

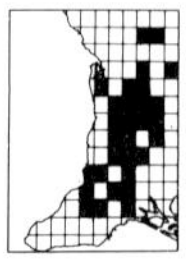

11. *Paspalidium constrictum*, **knotty-butt paspalidium**: 5–80 cm tall; infl 6–11 cm long; spikelets (11a) c. 2.5 mm long. NW, LE, NU, GT, FR, EA, EP, NL, MU, YP; all mainland States. *P. jubiflorum*, **Warrego summer-grass**, has the spikelets regularly arranged along the infl branches.

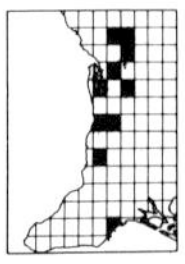

12. *Themeda triandra*, **kangaroo grass**: To 90 cm tall; infl 10–20 cm long; fertile spikelets 5–6 mm long. The figure (12a) shows a pair of large bracts enclosing 1 fertile and 4 sterile or male spikelets. NW, LE, GT, FR, EA, EP, NL, MU, YP, SE; all States; Africa and Asia.

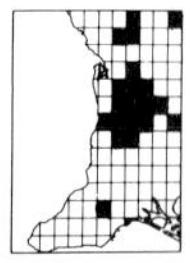

(Continued)

1a
1
2a
2
3
3a
4a
4
5
5a
6a
6
7b
7
7a
8a
8
9a
9
10a
10
11a
11
12a
12

Plate LXXXIV

GRAMINEAE
(Continued)

1. *Aristida behriana*, **brush-wire-grass**: 24–40 cm tall; infl 6–15 cm long; spikelets (1a) 14–20 mm long excl. awns; note the 3-branched awn. FR, EA, EP, NL, MU, YP; Vic. 4 other species.

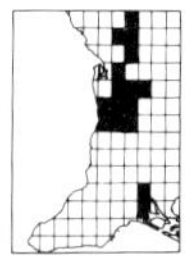

2. **Avellinia michelii*, **avellinia**: 5–30 cm tall; infl to 10 cm long; spikelets (2a) c. 5 mm long; note the small first glume and the awns. EP, NL, MU, YP, KI, SE; WA, Vic.; native to the Mediterranean.

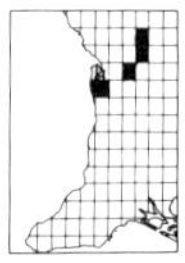

3. **Bromus catharticus*, **prairie grass**: 40–100 cm tall; infl to 35 cm long; spikelets (3a) 35–40 mm long; the only perennial. LE, FR, EP, NL, MU, YP, SE; all States except NT; native to S. America.

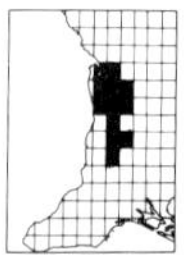

4. **Bromus diandrus*, **jabbers**: 30–90 cm tall; infl to 25 cm; spikelets (4a) to 35 mm long; some infl branches usually more than 2 cm long. LE, NU, GT, FR, EP, NL, MU, YP, KI, SE; all States; native to the Mediterranean region and south-west Europe. **Bromus rigidus*, rigid brome, has infl branches less than 2 cm long.

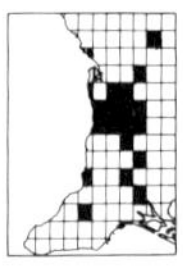

5. **Bromus hordeaceus*, **soft brome**: 55–80 cm tall; infl 50–85 mm long; spikelets (5a) 12–18 mm long. EP, NL, MU, YP, KI, SE; all States except NT; native to Europe.

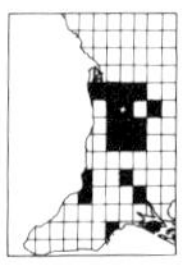

6. **Bromus madritensis*, **Madrid brome**: 18–60 cm tall; infl 4–11 cm long; spikelets (6a) 25–35 mm long; stem lacking hairs below infl FR, EP, NL, MU, YP, KI, SE; all States except NT; native to southern parts of Europe.

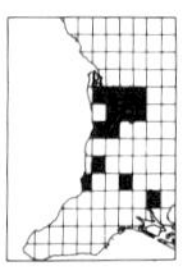

7. **Bromus rubens*, **red brome**: 15–40 cm tall; infl 3.5–7.0 cm long; spikelets (7a) 25–35 mm long; stem with hairs below infl The red colour only develops with maturity. NU, FR, EA, EP, NL, MU, YP, ?KI, SE; all mainland States except NT; native to southern Europe. 2 other species.

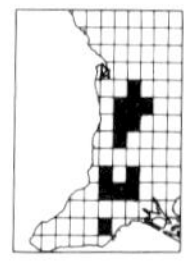

8. **Dactylis glomerata*, **cocksfoot**: 30–100 cm tall; infl 5–30 cm long; spikelets (8a) c. 7 mm long. YP, SE; all States except NT; native to Europe, N. Africa and Asia.

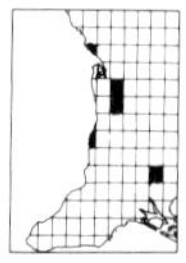

9. *Glyceria australis*, **Australian sweet grass**: 40–100 cm tall; infl 30–55 cm long; spikelets (9a) 1–3 cm long; lemmas rounded along their backs. Grows in or near freshwater of creeks and swamps. FR, MU, KI, SE; all States except NT and Qld.

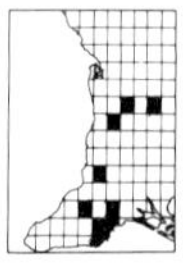

10. **Pentaschistis airoides*, **false hair-grass**: Annual 5–16 cm tall; infl 2–4 cm long; spikelets (10a) c. 3 mm long excl. awns. FR, EP, MU, SE; WA, NSW, Vic.; native to southern Africa. Differs from species of *Aira* in having hairs.

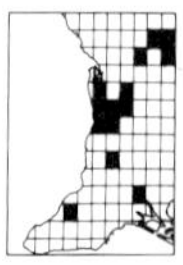

11. **Pentaschistis thunbergii*, **pussy tail**: Perennial c. 30 cm tall; infl 2–5 cm long; spikelets (11a) 4–5 mm long excl. the awns; perennial. NL, MU, SE; WA; native to southern Africa.

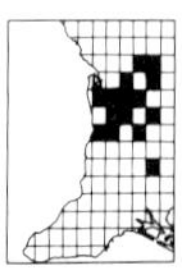

12. **Stenotaphrum secundatum*, **buffalo grass**: Creeping lawn grass sometimes to 70 cm tall; infl 4–8 cm long; spikelets (12a) 4–5 mm long. MU, YP, KI, SE; all States except NT; first described from N. America.

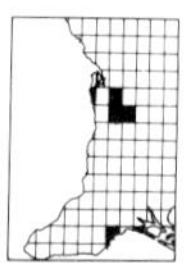

(Continued)

1a
1
2
2a
3a
3
4a
4
5a
5
6a
6
7
7a
8a
8
9
9a
10a
10
11
11a
12a
12

Plate LXXXV

GRAMINEAE
(*Continued*)

1. *Danthonia auriculata*, **lobed wallaby-grass**: 15–50 cm tall; infl 2–5 cm long; spikelets 9–16 mm long excl. awns; note the 'ears' on the lemma (1a) edges. FR, NL, MU; NSW, Vic.; introduced into New Zealand.

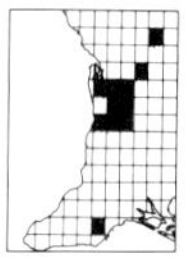

2. *Danthonia caespitosa*, **white top**: 20–90 cm tall; infl 2–12 cm long; spikelets 14–25 mm long excl. awns; note 3 dense rows of tufts on the lemma (2a) which is 3–6 mm long from tip to notch; hairs round leaf bases where they meet the stem 3–4 mm long. NW, NU, GT, FR, EA, EP, NL, MU, YP, KI, SE; all States except NT and Qld.

3. *Danthonia clelandii*, a **wallaby-grass**: To 75 cm tall; infl 8–13 cm long; spikelets 16–21 mm long excl. awns; scattered hairs on the lemma (3a) as well as distinct tufts. Restricted to this region.

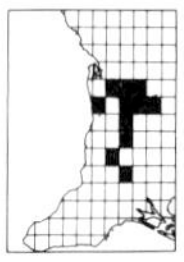

4. *Danthonia geniculata*, **kneed wallaby-grass**: 10–45 cm tall; infl 1.5–5.0 cm long; spikelets 5–8 mm long excl. awns; note the short central awn of the lemma (4a) and awns on side lobes 0. FR, MU, ?YP, KI, SE; Vic.

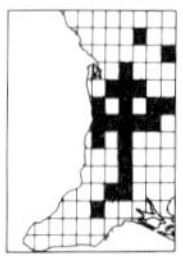

5. *Danthonia pilosa*, **velvet wallaby-grass**: To 90 cm tall; infl 4–7 cm long; spikelets 9–20 mm long excl. awns; lemma (5a). FR, ?EP, NL, MU, KI, SE; all States except NT; New Zealand. Hard to separate from *D. caespitosa*, **white top**.

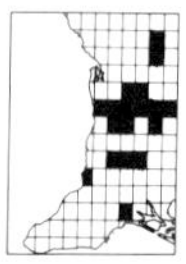

6. *Danthonia racemosa*, a **wallaby-grass**: 20–75 cm tall; infl 5–15 cm long; spikelets c. 15 mm long excl. awns; note the gap in the uppermost row of hair tufts on the lemma (6a) and the long narrow infl FR, EP, NL, MU, KI, SE; all States except NT and Tas; New Zealand.

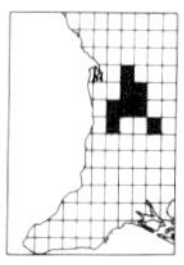

7. *Danthonia setacea*, **bristly wallaby-grass**: 10–60 cm tall; infl 3–5 cm long; spikelets 8–14 mm long excl. awns; lemma (7a). NU, FR, EP, NL, MU, YP, KI, SE; all States except NT and Qld. Similar to *D. caespitosa*, **white top**, but lemmas to 3 mm long to the notch and hairs round leaf bases to 2 mm long.

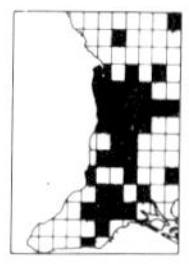

8. *Danthonia tenuior*, a **wallaby-grass**: 10–90 cm tall; infl 5–10 cm long; spikelets 10–17 mm long excl. awns; lemma (8a). ?NL, SE; all States except WA and NT; introduced into New Zealand. Hard to distinguish from *D. setacea*, **bristly wallaby-grass**. 6 or 7 other species.

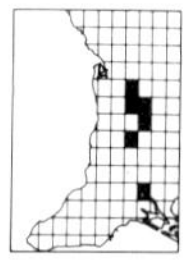

9. **Gastridium phleoides*, **nit-grass**: To 50 cm tall; infl 4–10 cm long; spikelets (9a) 5–6 mm long excl. awns; note the swollen spikelet base. FR, EP, NL, MU, SE; NSW, Vic.; native to Europe and N. Africa.

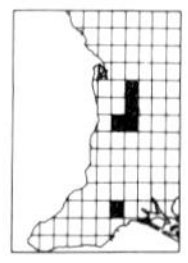

10. **Pennisetum villosum*, **feather-top**: 40–50 cm tall; infl 4–8 cm long; spikelets (10a) 10–12 mm long but concealed by longer feathery bristles. FR, EP, NL, MU, YP, KI, SE; all States except NT and Tas. Looks rather like *Aristida behriana*, **bush wire-grass**, which has 3-branched awns. 3 other species incl. the common lawn grass **P. clandestinum*, kikuyu.

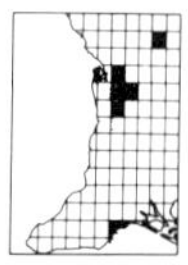

11. *Phragmites australis*, **common reed**: 1–3 m tall; infl 12–40 cm long; spikelets (11a) 10–18 mm long. The large reed common next to water. LE, FR, EA, NL, MU, SE; all mainland States; cosmopolitan.

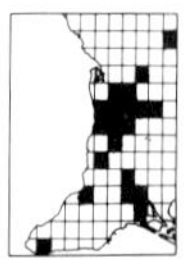

12. **Setaria verticillata*, **whorled pigeon-grass**: Occasionally to 1 m tall; infl to 12 cm long; spikelets (12a) c. 2 mm long. A common weed whose infl easily sticks to clothing and breaks up. LE, FR, EP, NL, MU, YP, KI, SE; all mainland States; native to Europe, Asia, Indonesia and the Americas. 4 other species.

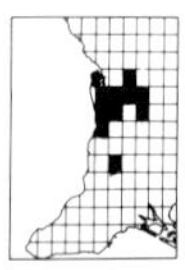

(*Continued*)

1
1a
2
2a
3
3a
4
4a
5
5a
6
6a
7
7a
8
8a
9
9a
10
10a
11
11a
12
12a

Plate LXXXVI

GRAMINEAE

(Continued)

1. *Amphipogon strictus*, **grey-beard grass**: To 50 cm tall; infl to 2 cm long; spikelets (1a) 9–12 mm long; lemma with 3 awned lobes. GT, FR, NL, MU, SE; WA and Vic. *A. caricinus* has infl 2–4 cm long.

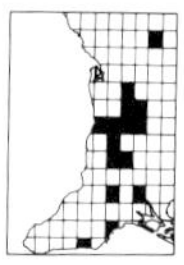

2. **Anthoxanthum odoratum*, **scented vernal grass**: 30–80 cm tall; infl 4–7 cm long; spikelets (2a and 2b) 7–8 mm long; note small hairy awned lemmas; plants sweetly scented. MU, KI, SE; all States except NT; native to Europe and Asia.

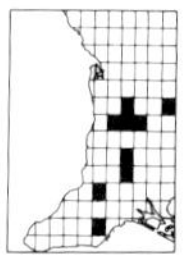

3. *Distichlis distichophylla*, **emu grass**: 10–30 cm tall; infl 2–5 cm long; spikelets (3a) 10–15 mm long. Common in wet saline areas. LE, EP, NL, MU, YP, KI, SE; NSW, Vic., Tas.

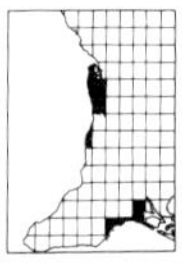

4. *Echinopogon ovatus*, **rough-bearded grass**: 15–70 cm tall; infl 2–4 cm long; spikelets (4a) 3.5–5.0 mm long. NU, FR, EP, NL, KI, SE; all States except NT. Like a prickly version of *Lagurus ovatus*, **hares tail-grass**.

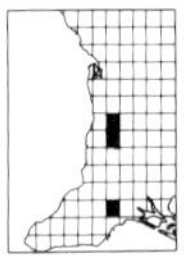

5. *Enneapogon nigricans*, **black-heads**: 30–40 cm tall; infl 3–7 cm long; spikelets 4–7 mm long excl. the awns; note the characteristic 9-awned lemmas (5a). NW, LE, GT, FR, EA, EP, NL, MU, YP, SE; all States except WA and Tas.

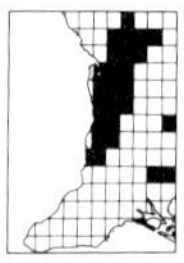

6. **Hordeum leporinum*, **barley-grass**: To 30 cm tall; infl (6a) 3–10 cm long; spikelets to 5 cm long incl. the long awns; spikelets coming away in threes (6b). LE, NU, GT, FR, EA, EP, NL, MU, YP, SE; South Neptune Is; all States; native to Europe and Asia. 4 other hard to recognise species including **H. vulgare*, **true barley**.

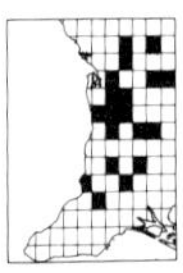

7. **Lagurus ovatus*, **hares tail grass**: 8–80 cm tall; infl 2–4 cm long; spikelets (7a) 8–10 mm long incl. the awns; a common white soft-headed grass especially near the sea. NU, EP, NL, MU, YP, KI, SE; all States except NT; native to the Mediterranean.

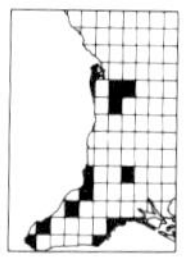

8. **Lophochloa cristata*, **annual cats tail**: 6–40 cm tall; infl 1–15 cm long; spikelets (8a) 4–5 mm long incl. the awns. A common weed. NU, FR, EA, EP, NL, MU, YP, KI, SE; all States except NT; native to the Mediterranean. **L. pumila*, **tiny bristle-grass**, has softly hairy glumes.

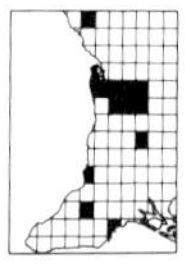

9. **Phalaris aquatica*, **phalaris**: To 50 cm tall; infl 5–12 cm long; spikelets (9a) 6–7 mm long; the only perennial species. NL, SE; all mainland States except NT; native to the Mediterranean. Cultivated and naturalised in many countries.

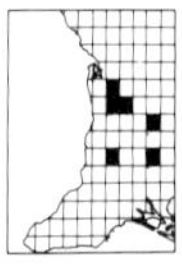

10. **Phalaris minor*, **lesser canary-grass**: To 50 cm or more tall; infl 2–6 cm long; spikelets (10a) 5–6 mm long. LE, FR, EA, EP, NL, MU, YP, KI, SE; all States except NT; native to the Mediterranean. 2 other annual species.

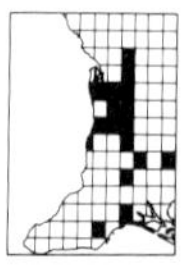

11. **Polypogon monspeliensis*, **annual beard-grass**: 4–50 cm tall; infl 1–12 cm long; spikelets (11a) 2.0–2.5 mm long. LE, FR, EA, EP, NL, MU, YP, KI, SE; all States except NT; New Zealand, the Americas; native to Europe, Asia and N. Africa. **P. maritimus*, **coast beard-grass**, has more pronounced pointed lobes below the awns.

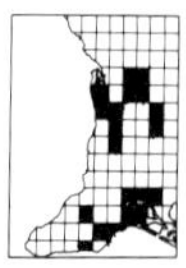

12. **Polypogon viridis*, **water bent**: 10–100 cm tall; infl 4–10 cm long; spikelets (12a) c. 2 mm long. FR, EA, EP, NL, SE; WA, NSW, Vic.; native to the Mediterranean and Asia. The only perennial species.

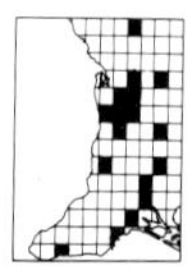

(Continued)

1a
1
2a
2b
2
3a
3
4a
4
5a
5
6a
6
6b
7a
7
8a
8
9a
9
10a
10
11a
11
12a
12

Plate LXXXVII

GRAMINEAE
(*Continued*)

1. *Ammophila arenaria*, **marram grass**: 60–120 cm tall; infl 7–25 cm long; spikelets (1a) 12–15 mm long. Introduced to stabilise coastal dunes. EP, NL, YP, KI, SE; all States except NT and Qld; native to Europe.

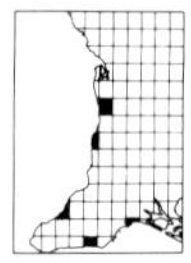

2. **Vulpia bromoides*, **squirrel-tail grass**: To 50 cm tall; infl 2–10 cm long; spikelets (2a) 6–10 mm long excl. awns; lemma (2b). NU, FR, EP, NL, YP, MU, KI, SE; all States except NT; native to the Mediterranean and western Europe.

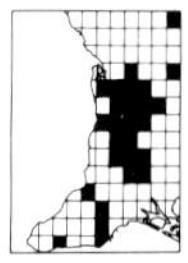

3. **Vulpia megalura*, **fox-tail fescue**: Usually less than 40 cm tall; infl 5–20 cm long; spikelets (3a) 10–15 mm long excl. awns; note the long hairs along the lemma margins; lemma (3b). FR, EP, NL, MU, YP, KI, SE; all States except NT; native to the Mediterranean, early introduction to N. America.

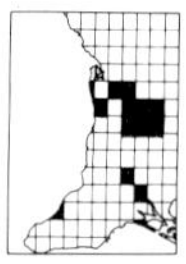

4. **Vulpia myuros*, **rats-tail fescue**: To 50 cm tall; infl 5–30 cm long; spikelets (4a) 7–10 mm long excl. awns; note the very short first glume; lemma (4b). LE, FR, EA, EP, NL, MU, YP, KI, SE; all States except NT; native to the Mediterranean and from central Europe to Great Britain. 3 other species.

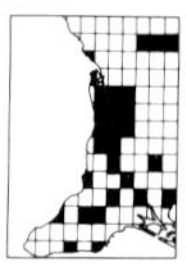

5. **Brachypodium distachyon*, **false brome**: 10–50 cm tall; infl 3–8 cm long; spikelets (5a) 2–3 cm long excl. awns. Note the long hairs along the lemma margins. FR, EP, NL, MU, YP, SE; all States except NT and Qld; native to the Mediterranean, Europe and Asia.

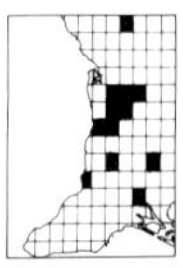

6. *Hemarthria uncinata*, **mat grass**: 20–80 cm tall; infl 6–14 cm long; spikelets (6a) c. 10 mm long. Note the hook often terminating the glumes. EP, MU, KI, SE; all States except NT.

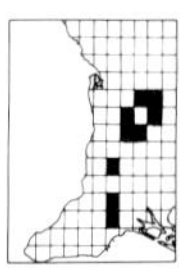

7. **Lolium multiflorum*, **Italian ryegrass**: To 1.2 m tall; infl 12–36 cm long; spikelets (7a) 4–8 mm long excl. awns. Restricted to this region; all States except NT; native to Europe.

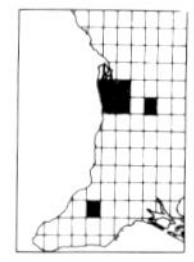

8. **Lolium perenne*, **perennial ryegrass**: To 90 cm tall; infl 4–20 cm long; spikelets (8a) 5–20 mm long; awns 0; note that the larger glume is not more than half the length of the spikelet. FR, NL, MU, YP, SE; all mainland States; native to Europe.

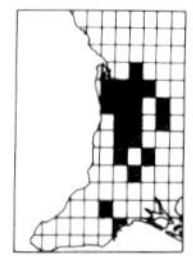

9. **Lolium rigidum*, **annual ryegrass**: To 1.2 m tall; infl 3–30 cm long; spikelets (9a) 4–18 mm long; awns 0; note that the larger glume is about as long as the spikelet. LE, NU, GT, FR, EA, EP, NL, MU, YP, KI, SE; all States; native to the Mediterranean. 2 other species

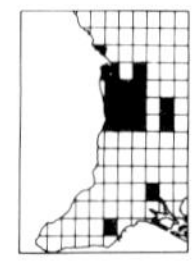

10. *Puccinellia stricta*, **Australian saltmarsh-grass**: 15–60 cm tall; infl 5–15 cm long; spikelets (10a) 6–10 mm long; lemmas rounded along the backs. EP, NL, MU, YP, KI, SE; all States except NT and Qld. **P. distans*, **reflexed poa**, usually has smaller spikelets.

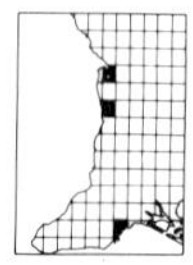

11. *Sporobolus virginicus*, **salt couch**: 7–50 cm tall; infl 1–6 cm long; spikelets (11a) c. 3 mm long. LE, NU, FR, EP, NL, MU, YP, KI, SE; all mainland States; all temperate continents. In growth often rather like *Distichlis distichophylla*, **emu grass**. Usually grows in wettish saline areas. 3 other species.

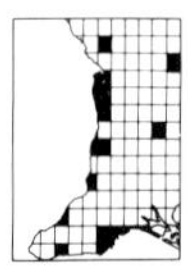

12. **Parapholis incurva*, **curly ryegrass**: 5–30 cm tall; infl 3–10 cm long; spikelets (12a) c. 6 mm long. LE, NU, FR, EP, NL, MU, YP, KI, SE; all States except NT and Qld; native to Europe, N. Africa and Asia. Note the 2 glumes both facing out from the stem. Similar to *Lolium* but with a curved infl Also some similarity with *Hemarthria* and several other genera not included in this book.

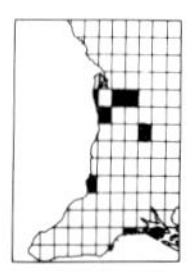

1
1a
2
2a
2b
3
3a
3b
4
4a
4b
5
5a
6
6a
7
7a
8
8a
9
9a
10
10a
11
11a
12
12a

PLATE LXXXVIII

TYPHACEAE

1. *Typha domingensis*, a **bulrush**: To 2 m or more tall; infl 25–55 cm long, the upper male part separated from the female part by a space of 2–5 cm. Common in wet places. NW, LE, FR, MU, YP, KI, SE; all States; Africa, Asia. *T. orientalis*, another **bulrush**, has a thicker infl (15–20 mm thick), the female less than 2 cm from the male.

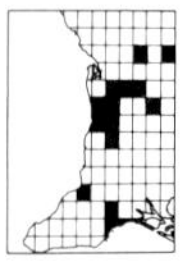

CYPERACEAE

2. *Carex appressa*, **tall sedge**: To 1 m or more tall; stems (2a) 1–3 mm broad; bladder surrounding ovary (2b) 2–3 mm long; infl branches stalkless, usually more numerous than shown. FR, MU, KI, SE; all States except NT; western Pacific.

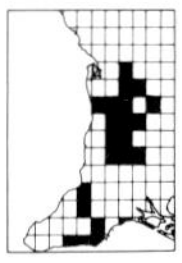

3. *Carex bichenoviana*, a **sedge**: 25–50 cm tall; stems (3a) 1–2 mm broad; bladder surrounding ovary (3b) 4.5–5.0 mm long. FR, MU; NSW, Vic., Tas.

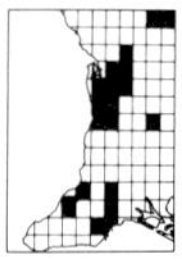

4. *Carex breviculmis*, a **sedge**: 15–40 cm tall; stems (4a) c. 0.5 mm broad; bladder surrounding ovary (4b) 3–4 mm long, note hairs on it. FR, EP, NL, MU, KI, SE; all States except WA and NT.

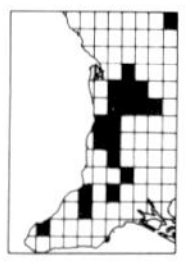

5. *Carex fascicularis*, **tassel sedge**: To over 50 cm tall; stems (5a) 1–2 mm broad; bladder surrounding ovary (5b) 4–5 mm long; lvs broader than most (5–10 mm). NL, MU, KI, SE; all States; western Pacific.

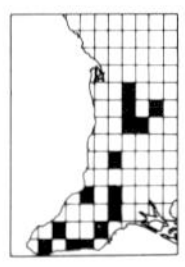

6. *Carex gaudichaudiana*, a **sedge**: To 60 cm tall or more; stems (6a) 0.5–1.0 mm broad; bladder surrounding ovary (6b) 2–3 mm long. FR, MU, KI, SE; all States except WA and NT.

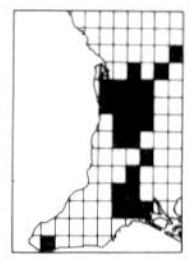

7. *Carex gunniana*, a **sedge**: 20–40 cm tall; stems (7a) 0.7–2.0 mm broad; bladder surrounding ovary (7b) 4.5–5.0 mm long. FR, SE; Vic. Tas.

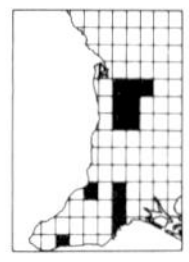

8. *Carex inversa*, **knob sedge**: 10–60 cm tall; stems (8a) 0.3–1.0 mm broad; bladder surrounding ovary (8b) 2.5–4.0 mm long. FR, EP, SE; all States except NT; New Zealand.

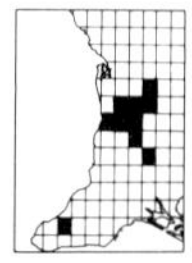

9. *Carex pumila*, **strand sedge**: 10–30 cm tall; stems (9a) 0.7–1.5 mm broad; bladder surrounding ovary (9b) c. 5 mm long. Infl branches shortly stalked. FR, NL; all States except NT; S. America and Asia.

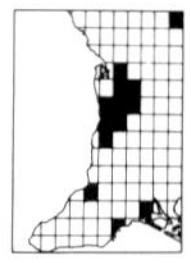

10. *Carex tereticaulis*, a **sedge**: To 1 m or more tall; stems (10a) 1.2–2.0 mm broad; bladder surrounding ovary (10b) 3.0–3.5 mm long. Infl branches usually more numerous than shown. FR, EP, NL, SE; all States except NT and Qld. 2 or 3 other species.

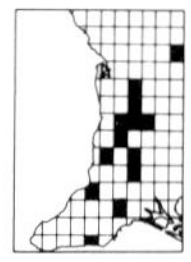

11. *Bolboschoenus caldwellii*, **sea club-rush**: 30–90 cm tall; fruit (11a) 3–4 mm long excl. style. LE, FR, EP, NL, MU, SE; WA, NSW, Vic., Tas.; New Zealand.

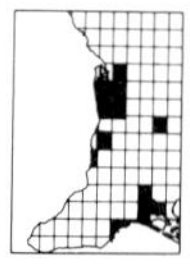

12. *Bolboschoenus medianus*, **marsh club-rush**: 70–200 cm tall; fruit (12a) 3.5–4.0 mm long excl. style which is 2- or 3-branched. LE, MU, SE; NSW, Vic., Tas.; New Zealand.

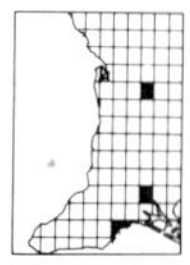

(Continued)

1
2a
2b
2
3a
3b
3
4a
4b
4
5a
5b
5
6a
6b
6
7a
7b
7
8a
8b
8
9a
9b
9
10a
10 b
10
11a
11
12a
12

Plate LXXXIX

CYPERACEAE

(Continued)

1. *Baumea acuta*, **pale twig-rush**: 15–30 cm tall; spikelets (1a) 4–5 mm long; fruit (1b) c. 2 mm long; leaves and stems flat. KI, SE; WA, NSW, Vic., Tas.

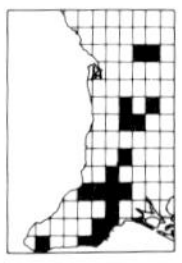

2. *Baumea arthrophylla*, a **twig-rush**: 1–2 m tall; spikelets (2a) 4.5–6.5 mm long; fruit (2b) 2.00–2.25 mm long; stems more or less compressed; hairs on spikelet bracts. LE, FR, KI, SE; all States; New Zealand.

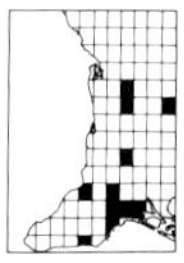

3. *Baumea articulata*, **jointed twig-rush**: 1–2 m tall; spikelets (3a) 4–5 mm long; fruit (3b) c. 2.5 mm long; note partitions in leaves and often stems. FR, NL, KI, SE; WA, Qld, NSW, Vic.; New Zealand, New Guinea.

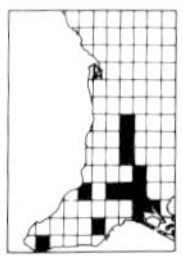

4. *Baumea gunnii*, **slender twig-rush**: 30–100 cm tall; spikelets (4a) 6–8 mm long; fruit (4b) 2.5–3.5 mm long; note furrowed stems. KI, SE; WA, Qld, NSW, Vic., Tas.; New Zealand, New Guinea.

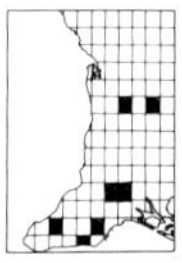

5. *Baumea juncea*, **bare twig-rush**: 30–100 cm tall; spikelets (5a) 4–5 mm long; fruit (5b) c. 3 mm long; leaves reduced to points. LE, EP, MU, YP, KI, SE; all States except NT; New Zealand.

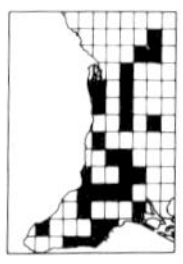

6. *Baumea laxa*, **lax twig-rush**: 15–60 cm tall; spikelets (6a) 4–5 mm long; fruit (6b) 2.0–2.5 mm long; stems and leaves flat. KI, SE; WA, Vic.

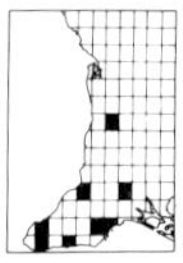

7. *Baumea rubiginosa*, **soft twig-rush**: 30–100 cm tall; spikelets (7a) 5–8 mm long; fruit (7b) 3–4 mm long; stems more or less cylindrical. LE, FR, KI, ?SE; all States; New Zealand, Asia.

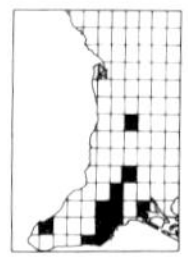

8. *Baumea tetragona*, **square twig-rush**: 30–120 cm tall; spikelets (8a) 3.5–4.0 mm long; fruit (8b) c. 2 mm long; note 4-angled stems. KI, SE; all States except WA and NT.

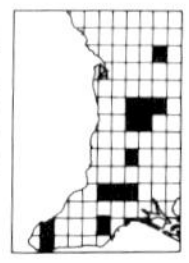

9. *Schoenoplectus litoralis:* To 1.5 m tall; spikelets (9a) 8–15 mm long; fruit c. 2 mm long; stem cylindrical; feathery bristles round fruit usually 4. LE, FR, EA, NL, MU, SE; WA, NT, Qld, NSW; Africa, Europe, Asia.

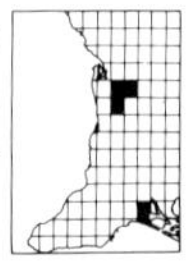

10. *Schoenoplectus pungens*, **American club-rush**: 30–60 cm tall; spikelets (10a) 6–10 mm long; fruit c. 3 mm long; stems 3-angled. LE, GT, FR, EP, MU, SE; WA, Vic.; New Zealand, Europe, the Americas.

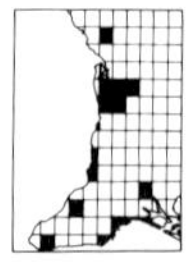

11. *Schoenoplectus validus*, **river club-rush**: To 3 m or more tall; spikelets (11a) 5–10 mm long; fruit 2.0–2.3 mm long; stem cylindrical; feathery bristles round fruit 5 or 6. LE, FR, EP, NL, MU, KI, SE; all States except NT; cosmopolitan.

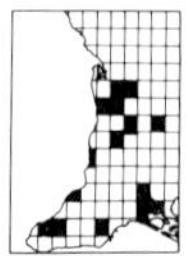

12. *Tetraria capillaris*, **hair sedge**: 20–40 cm tall; spikelets (12a) 3.5–5.0 mm long; fruit 3–4 mm long incl. style base. KI, SE; all States except NT; New Zealand.

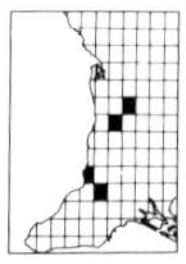

(Continued)

1a
2a
3a
1b
2b
3b
3
4a
4b
4
5a
5b
5
6a
6b
6
1
2
7a
7b
7
8a
8b
8
9a
9
10b
10a
10
11a
11
12a
12

Plate XC

CYPERACEAE
(Continued)

1. ******Cyperus congestus*, **dense flat-sedge**: 20–60 cm tall; spikelets (1a) 8–25 mm long, bracts c. 4 mm long. SE; WA, NSW, Vic.; native to S. Africa.

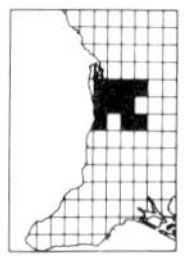

2. *Cyperus gymnocaulos*, **spiny flat-sedge**: 15–70 cm tall; spikelets (2a) 3.0–5.5 mm long; note reduced leaves. NW, LE, GT, FR, EA, EP, NL, MU, SE; all mainland States.

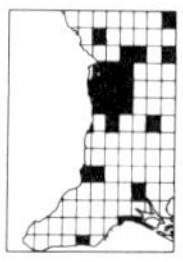

3. **Cyperus rotundus*, **nut-grass sedge**: 15–30 cm tall; spikelets (3a) 10–20 mm long, bracts 3.0–3.5 mm long. GT, MU; all mainland States; native to warmer parts of the world. A troublesome garden weed.

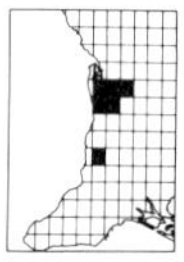

4. *Cyperus tenellus*, tiny **flat-sedge**: To 8 cm tall; spikelets (4a) 4–9 mm long. EP, KI, SE; all States except NT and Qld; New Zealand; native to S. Africa and possibly naturalised in Australia.

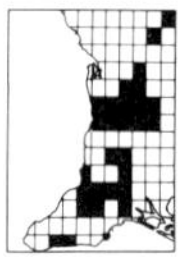

5. *Cyperus vaginatus*, a flat-sedge: 30–150 cm tall; spikelets (5a) 8–18 mm long; note reduced leaves. NW, LE, GT, FR, EA, EP, NL, MU, YP, KI, ?SE; all mainland States. 7 or 8 other species.

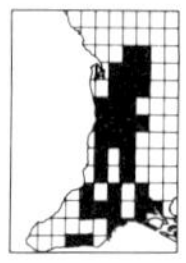

6. *Schoenus apogon*, **common bog-rush**: Mostly 5–25 cm tall; spikelets (6a) 4–7 mm long; fruit (6b) c. 1 mm long; note long bracts. FR, EP, NL, MU, YP, KI, SE; all States except WA and NT; New Zealand, New Guinea and ?Japan.

7. *Schoenus brachyphyllus:* 30–60 cm tall; spikelets 8–10 mm long; fruit (7a) c. 1.7 mm long; spikelet bracts woolly-ciliate on the margins. KI, SE.

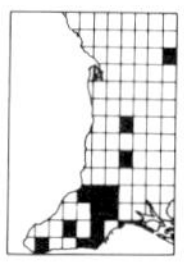

8. *Schoenus breviculmis*, **matted bog-rush**: Often c. 5 cm tall, with the infl shorter than the leaves; spikelets c. 10 mm long; fruit (8a) c. 1.1 mm long. EP, MU, YP, KI, SE; WA, Vic.

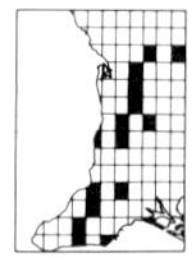

9. *Schoenus carsei:* 30–75 cm tall; spikelets 5–9 mm long; fruit (9a) c. 1.5 mm long; spikelet bracts hairless. EP, KI, SE; Vic.; New Zealand.

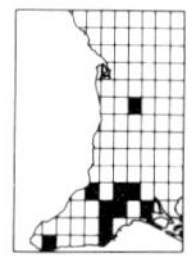

10. *Schoenus maschalinus*, **leafy bog-rush**: To 20 cm tall; spikelets 2–3 mm long; fruit (10a) c. 1 mm long. KI, SE; all States except NT; western Pacific.

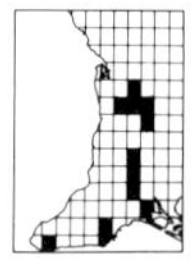

11. *Schoenus nitens*, **shiny bog-rush**: 3–25 cm tall; spikelets 3–5 mm long, arranged in a single head; fruit (11a) 1.7–2.0 mm long. NL, MU, YP, KI, SE; all States except NT; New Zealand, New Guinea, S. America. 7 other species.

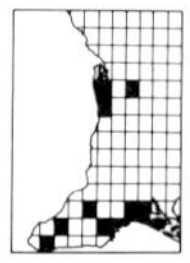

12. *Cladium mariscus*, **leafy twig-rush**: 1–2 m tall; spikelets 2.5–5.0 mm long; fruit 2.0–2.5 mm long. NL, SE; all States except ?Tas.; tropical and temperate parts of the world.

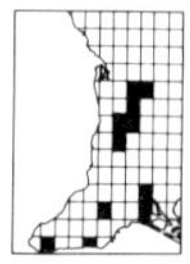

(Continued)

1a
1
2a
2
3a
3
4a
4
5a
5
6a
6b
6
7a
7
8a
8
9a
9
10a
10
11a
11
12

Plate XCI

CYPERACEAE

(Continued)

1. *Isolepis cernua*, **nodding club-rush**: To 10 cm or rarely more tall; spikelets 2.5–3.0 mm long; fruit (1a) c. 0.9 mm long; spikelet bracts more rounded at the tip than in *I. marginata*. FR, EP, NL, MU, YP, KI, SE; all States except NT; most parts of the world.

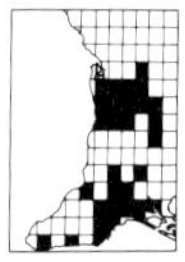

2. *Isolepis fluitans*, **floating club-rush**: Creeping or floating; spikelets 3–5 mm long; fruit (2a) 1.0–1.5 mm long; note style is 2-branched. EP, MU, KI, SE; all States except NT; Europe, Africa, Asia.

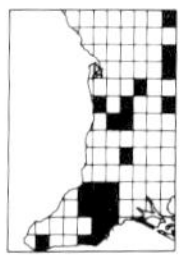

3. *Isolepis hookerana*: To 12 cm tall; spikelets 2.5–3.5 mm long; fruit (3a) c. 0.6 mm long; note pattern of fruit surface. NW, LE, FR, EP, KI, SE; all States except NT.

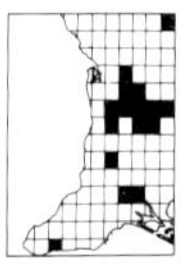

4. *Isolepis inundata*, **swamp club-rush**: 10–30 cm tall; spikelets 4–6 mm long; fruit (4a) c. 1 mm long; note style is 2- or 3-branched; only 1 or 2 stamens. LE, FR, EP, NL, MU, KI, SE; all States except NT; New Zealand, S. America, Asia.

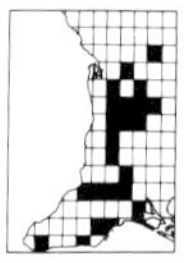

5. *Isolepis marginata*, **coarse club-rush**: Usually less than 8 cm tall; spikelets 3–4 mm long; fruit (5a) 0.9–1.0 mm long. FR, EP, NL, MU, YP, KI, SE; all States except NT, Qld and Tas. New Zealand, S. Africa.

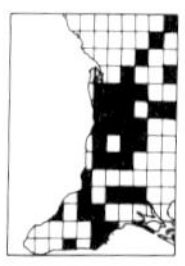

6. *Isolepis nodosa*, **knobby club-rush**: 50–150 cm tall; spikelets 4–5 mm long; fruit (6a) 0.9–1.0 mm long; note the very dense head. NU, FR, EP, MU, YP, KI, SE; all States except NT; New Zealand, S. America and S. Africa.

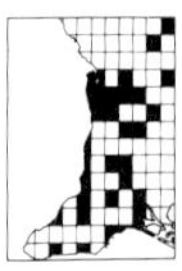

7. *Isolepis platycarpa*: Usually less than 8 cm tall; spikelets 2.5–3.5 mm long; fruit (7a) 0.7–0.8 mm long; differs from *I. cernua* and *I. hookerana* in fruit size and surface characters. FR, EP, MU, YP, KI, SE; all States except NT and Qld; New Zealand. 4 other species.

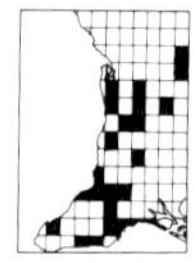

8. *Gahnia ancistrophylla*: 30–80 cm tall; spikelets (8a) c. 3.5 mm long, 2-flowered; fruit 1.8–2.0 mm long; note curled leaf tips. FR, EP, NL; WA, Vic.

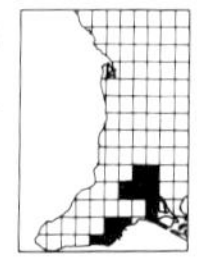

9. *Gahnia deusta*: 30–50 cm tall; spikelets (9a) 6–15 mm long, 2-flowered; fruit 1.8–2.5 mm long; leaves very roughly hairy in the upper half. EP, MU, YP, KI, SE; WA, Vic.

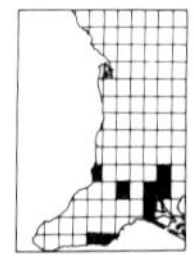

10. *Gahnia filum*, **chaffy saw-sedge**: 40–150 cm tall; spikelets (10a) 6–7 mm long, 1-flowered; fruit 5–6 mm long. EP, YP, SE; WA, NSW, Vic., Tas.

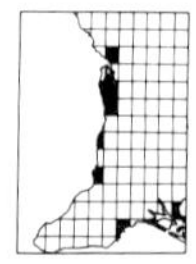

11. *Gahnia lanigera*, **desert saw-sedge**: 10–30 cm tall; spikelets (11a) 3–4 mm long, 2-flowered; fruit 1.6–1.9 mm long; leaf sheaths bearded at mouth. EP, NL, MU, YP, SE; WA, NSW, Vic.

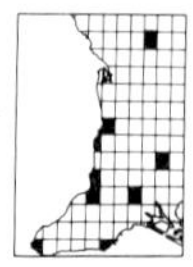

12. *Gahnia trifida*, **cutting grass**: Usually over 1 m tall; spikelets (12a) 4–5 mm long, 1-flowered; fruit c. 2 mm long; lower leaf-like bracts (12b) with small spine-like hairs. LE, EP, NL, MU, KI, SE; WA, Vic., Tas. 3 other species.

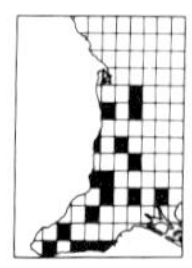

(Continued)

1a
2a
3a
4a
5a
5a
6a
6a
1
2
3
4
5
6
7a
8a
9a
10a
11a
12a
12b
7
8
9
10
11
12

Plate XCII

CYPERACEAE
(Continued)

1. *Eleocharis acuta*, **common spike-rush**: To 90 cm tall; spikelet 15–30 mm long; fruit (1a) 1.5–1.8 mm long. LE, GT, FR, EP, NL, MU, KI, SE; all States except NT; New Zealand.

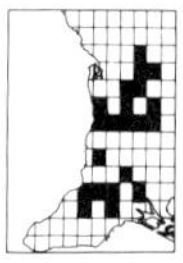

2. *Eleocharis gracilis*, **slender spike-rush**: To c. 20 cm tall; spikelet mostly 5–9 mm long; fruit (2a) 1.2–1.3 mm long. KI, SE; all States except NT; New Zealand. 3 or 4 other species.

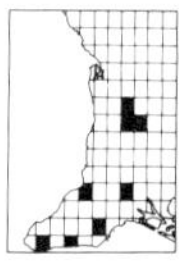

3. *Lepidosperma canescens*, **hoary rapier-sedge**: 25–100 cm tall; spikelets (3a) 5–7 mm long; fruit 2–3 mm long; stem (3b) cylindrical. NW, EP, MU, KI, SE; WA, Vic.

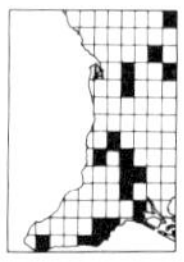

4. *Lepidosperma carphoides*, **black rapier-sedge**: 20–40 cm tall; spikelets (4a) 8–14 mm long, black; fruit c. 4 mm long; stems (4b) cylindrical. EP, MU, KI, SE; WA, Vic.

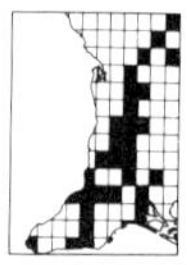

5. *Lepidosperma concavum*, **sand-hill sword-sedge**: To 60 cm tall; spikelets (5a) 5–6 mm long; fruit c. 2 mm long; stems (5b) flat or concave. EP, NL, MU, YP, KI, SE; Qld, NSW, Vic., Tas.

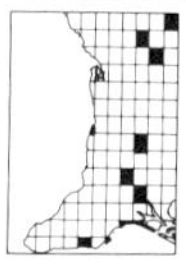

6. *Lepidosperma congestum*, **clustered sword-sedge**: 15–50 cm tall; spikelets (6a) 4–5 mm long; fruit c. 2.5 mm long; stems (6b) flat or convex on one side. EP, MU, YP, KI, SE; Vic.

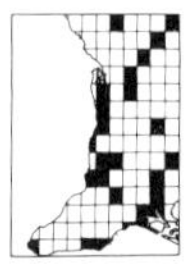

7. *Lepidosperma gladiatum*, **sword rush**: Over 1 m tall; spikelets (7a) 6–8 mm long; fruit c. 3 mm long; stems (7b) biconvex with sharp edges. EP, YP, KI, SE; all States except NT and Qld.

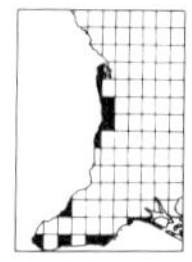

8. *Lepidosperma laterale*, **variable sword-sedge**: 40–100 cm tall; spikelets (8a) 4–8 mm long; fruit c. 4 mm long; stems (8b) flat or concave on one side. LE, FR, EP, NL, MU, YP, KI, SE; all States except WA and NT; New Zealand.

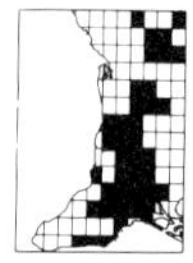

9. *Lepidosperma lineare*, **little sword-sedge**: 5–30 cm tall; spikelets (9a) 5–7 mm long; fruit 3.0–3.5 mm long; stems (9b) slightly biconvex. FR, NL, SE; NSW, Vic., Tas.

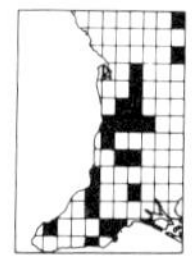

10. *Lepidosperma longitudinale*, **pithy sword-sedge**: 60–100 cm tall; spikelets (10a) 6–7 mm long; fruit c. 3 mm long; stems (10b) strongly biconvex. EP, KI, SE; all States except NT.

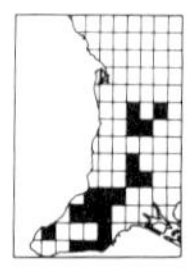

11. *Lepidosperma semiteres*, **wire rapier-sedge**: 30–100 cm tall; spikelets (11a) 6–8 mm long; fruit c. 4 mm long; stems (11b) angular or strongly biconvex. MU, KI, SE; NSW, Vic.

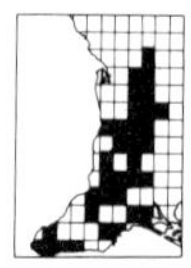

12. *Lepidosperma viscidum*, **sticky sword-sedge**: 20–50 cm tall; spikelets (12a) 4–6 mm long; fruit 2.5–3.0 mm long; stems flat or convex on one side, more or less viscid. FR, EP, NL, MU, YP, KI, SE; WA, NSW, Vic.

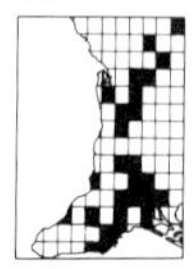

1
1a
2
2a
3
3a
3b
4
4a
4b
5
5a
5b
6
6a
6b
7
7a
7b
8
8a
8b
9
9a
9b
10
10a
10b
11
11a
11b
12
12a
12b

Plate XCIII

ORCHIDACEAE

1. *Caleana major*, **large duck-orchid**: 15–30 cm tall; lateral sepals 'duck's wings' c. 12 mm long; note the resemblance of the flower to a duck. SE; Vic. *Paracaleana minor*, **small duck-orchid**, has narrower leaves and the labellum 'duck's head' is covered by tubercles.

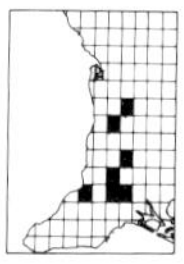

2. *Calochilus robertsonii*, **common beard-orchid**: Rarely above 45 cm tall; labellum 'bearded tongue' c. 20 mm long, the lower part covered by glands. KI, SE; all States except NT. 4 other species.

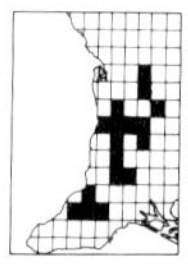

3. *Leporella fimbriata*, **fringed hare-orchid**: 15–20 cm tall; petal (3a); the conspicuously fringed labellum (3b) c. 9 mm broad. EP, NL, KI, SE; WA, Vic., Tas.

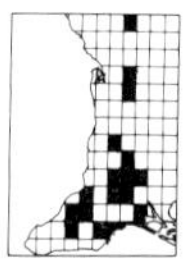

4. *Pterostylis biseta*, **rusty-hood**: Usually 10–35 cm tall; hood 1–3 cm long; labellum (4a) membranous, c. 6 mm long; note lvs mostly basal. FR, EA, EP, NL, MU, YP, SE; NSW, Vic.

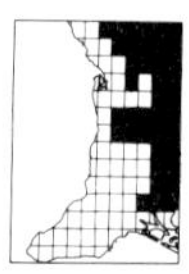

5. *Pterostylis curta*, **blunt green-hood**: 10–20 cm tall; labellum (5a) twisted at tip, c. 14 mm long; column (5b) c. 2 cm long; note lvs mostly basal. FR, SE; all States except WA and NT.

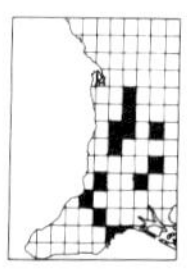

6. *Pterostylis longifolia*, **tall green-hood**: 10–80 cm tall; labellum (6a) with 2 small side lobes, c. 8 mm long; note lvs along stem. FR, EP, NL, MU, YP, KI, SE; all States except WA and NT.

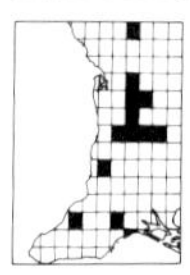

7. *Pterostylis nana*, **dwarf green-hood**: Usually 5–10 cm tall; labellum (7a) 4–5 mm long; column (7b) c. 8 mm long. Somewhat like *P. pedunculata* but differs in colour. GT, FR, EP, NL, MU, YP, KI, SE; all States except NT and Qld.

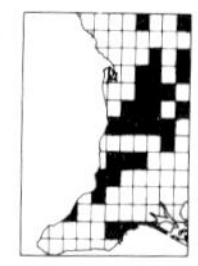

8. *Pterostylis nutans*, **nodding green-hood**: 7–20 cm tall; labellum (8a) c. 15 mm long; column (8b) c. 15 mm long; note fl facing down. KI, SE; all States except WA and NT; New Zealand.

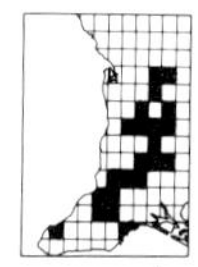

9. *Pterostylis pedunculata*, **maroon-hood**: 10–22 cm tall; labellum (9a) c. 5.5 mm long; column (9b) c. 11 mm long; see *P. nana*. FR, NL, YP, KI, SE; all States except WA and NT.

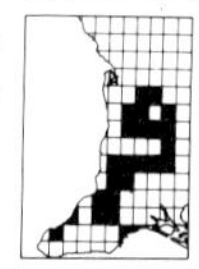

10. *Pterostylis plumosa*, **bearded green-hood**: 12–30 cm tall; labellum (10a) conspicuously hairy, c. 2 cm long. FR, EP, NL, MU, YP, KI, SE; all States except NT and Qld; New Zealand.

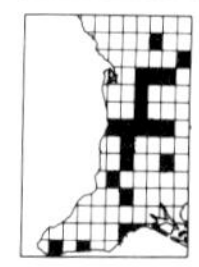

11. *Pterostylis robusta*, **larger striped green-hood**: 5–20 cm tall; labellum (11a) 8–10 cm long; column (11b) c. 16 cm long; note lvs along stems. FR, EP, NL, MU, YP, SE; NSW, Vic.

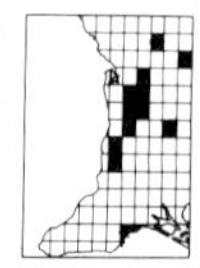

12. *Pterostylis sanguinea*, **banded green-hood**: 8–35 cm tall; labellum (12a) c. 5 mm long, with 2 lobes at the tip but not on the sides; note lvs along stems; closest to *P. longifolia*. FR, EP, NL, MU, YP, KI, SE; WA, Vic., Tas. 10 or 11 other species.

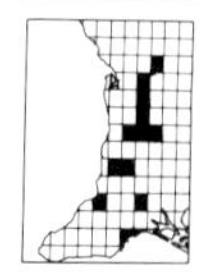

(*Continued*)

1
2
3
3a
3b
4
4a
5
5a
6
6a
7
7a
8
8a
9
9a
10
10a
11
11a
12
12a

Plate XCIV

ORCHIDACEAE

(Continued)

1. *Caladenia carnea*, **pink fingers**: 7–60 cm tall; 'petals' 11–20 mm long; 3-lobed labellum in the middle of the flower with 2 rows of glands. FR, EP, NL, MU, YP, KI, SE; all States except WA and NT; western Pacific to Malaysia.

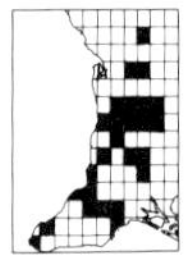

2. *Caladenia deformis*, **blue fairies**: 6–15 cm tall; labellum (2a) c. 11 mm long when stretched out, 3-lobed, with 4–6 rows of glands. FR, EP, NL, MU, YP, KI, SE; all States except NT and Qld.

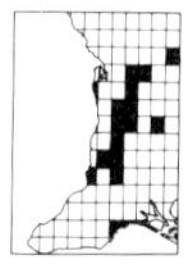

3. *Caladenia dilatata*, **king spider**: 10–50 cm tall; column (3a) note the 2 yellow glands at the base, 10–15 mm long. FR, EP, NL, MU, YP, KI, SE; all States except NT.

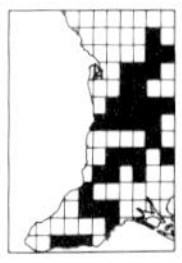

4. *Caladenia latifolia*, **pink fairies**: 15–30 cm tall; labellum (4a) 6–7 mm long, distinctly 3-lobed, with glands all near the base. FR, EP, NL, MU, YP, KI, SE; all States except NT and Qld.

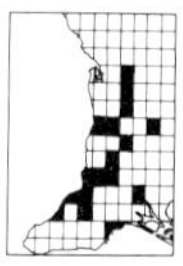

5. *Caladenia leptochila*, **narrow-lipped spider-orchid**: 15–45 cm tall; column (5a) 10–12 mm long; resembles *C. dilatata* in having the yellow glands but has an almost smooth-margined labellum. FR, NL, MU; Vic.

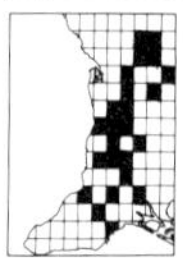

6. *Caladenia menziesii*, **rabbit orchid**: Usually 6–30 cm tall; lvs virtually lacking hairs; labellum (6a) c. 7 mm long. FR, EP, YP, KI, SE; all States except NT and Qld.

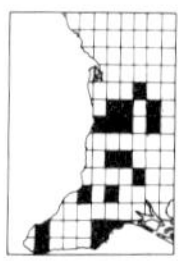

7. *Caladenia patersonii*, **white spider-orchid**: 20–50 cm tall; column (7a) 8–15 mm long; note fringed labellum. FR, EP, NL, MU, YP, KI, SE; all States except NT. 8 or 9 other species.

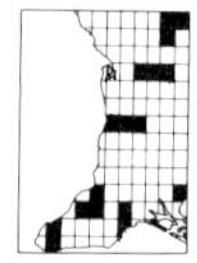

8. *Prasophyllum australe*, **austral leek-orchid**: 25–75 cm tall; 'petals' c. 8 mm long; ovary lying close to the stem. KI, SE; all States except WA and NT.

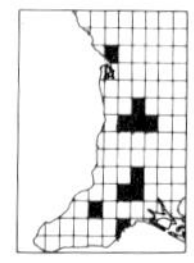

9. *Prasophyllum elatum*, **tall leek-orchid**: 45–150 cm tall; 'petals' c. 10.0–10.5 mm long; ovary lying close to the stem. FR, EP, NL, MU, YP, KI, SE; all States except NT.

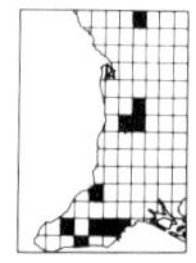

10. *Prasophyllum occultans*, **slaty leek-orchid**: 15–80 cm tall; 'petals' 11–13 mm long; column (10a) 2–3 mm long; ovary standing out from the stem. FR, EP, NL, MU, YP, SE; NSW, Vic., Tas.

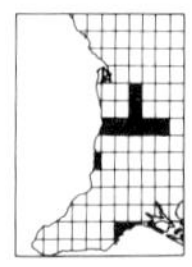

11. *Prasophyllum patens*, **scented leek-orchid**: 10–90 cm tall; 'petals' 5–8 mm long; ovary standing out from the stem. FR, EA, EP, NL, MU, YP, KI, SE; all States except NT; New Zealand.

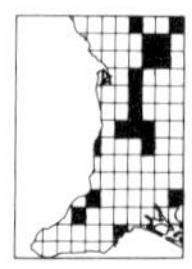

12. *Genoplesium rufum*, **red midge-orchid**: 4–30 cm tall; 'petals' 2–4 mm long; labellum (12a) c. 2.5 mm long; column (12b) c. 1 mm long; ovary standing out from the stem. FR, EP, NL, MU, YP, KI, SE; all States except WA and NT.

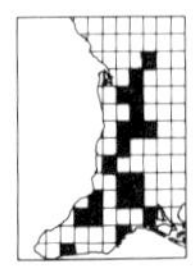

(Continued)

1
a
2a
2
3
3a
4
4a
5
5a
6
6a
7
7a
8
8a
9
9a
10
10a
11
11a
12
12a
12b

PLATE XCV

ORCHIDACEAE

(*Continued*)

1. *Acianthus caudatus*, **mayfly orchid**: 7-15 cm tall; lf broad, red below; upper 'petal' with a point 2.0-3.5 cm long. KI, SE; NSW, Vic., Tas.

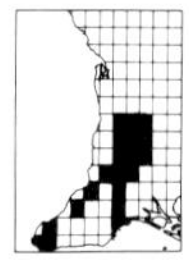

2. *Acianthus exsertus*, **mosquito orchid**: 7-16 cm tall; lf broad, red below; fls (2a); upper 'petal' 5-6 mm long incl. the short point. FR, EP, NL, MU, YP, KI, SE; NSW, Vic., Tas.

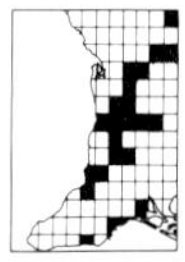

3. *Corybas diemenicus*, **veined helmet-orchid**: 1.5-4.0 cm tall; lvs almost circular; labellum (3a) fringed, the margins flared outwards. MU, KI, SE; ?WA, NSW, Vic., Tas. 4 other species.

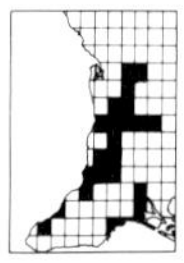

4. *Cyrtostylis robusta*, **large gnat orchid**: 7-15 cm tall; lf broad, green below; upper 'petal' 8-10 mm long. EP, NL, MU, YP, KI, SE; all States except NT; New Zealand.

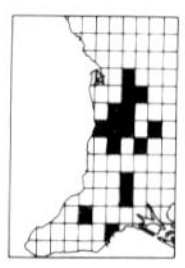

5. *Cryptostylis subulata*, **moose orchid**: 30-90 cm tall; fls (5a); the 5 'petals' c. 13-20 mm long. KI, SE; all States except WA and NT; New Zealand.

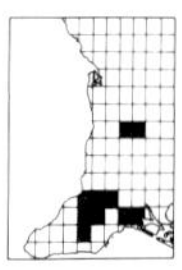

6. *Dipodium punctatum*, **hyacinth orchid**: 30-75 cm tall; lvs 0; fls pink or white, sometimes spotted; 'petals' c. 17 mm long. SE; all States except WA and NT.

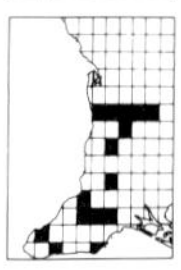

7. *Diuris lanceolata*, **cowslip**: Usually 20-40 cm tall; lvs narrow; fls mainly yellow; labellum (7a) c. 18 mm long. SE; NSW, Vic., Tas.

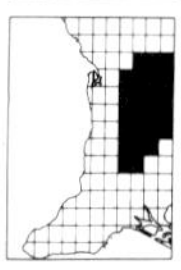

8. *Diuris corymbosa*, **bulldogs**: 10-45 cm tall; lvs narrow; fls yellow, brown and purple; labellum (8a) c. 10 mm long. EP, MU, YP, KI, SE; all States except NT and Qld.

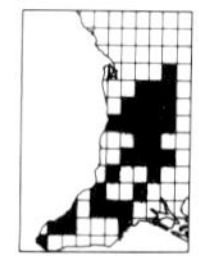

9. *Diuris maculata*, **common donkey-orchid**: Rarely over 30 cm tall; lvs narrow; fls yellow and dark brown; labellum (9a) c. 6 mm long; note the narrow side 'petals' which cross one another. EP, NL, MU, YP, SE; NSW, Vic., Tas. 3 other species.

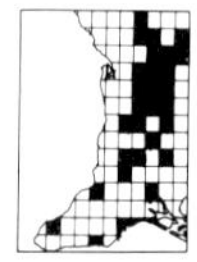

10. *Eriochilus cucullatus*, **parsons bands**: 7-22 cm tall; the large lower 'petals' white or pale pink, 12-17 mm long; labellum (10a) c. 6 mm long. FR, EP, NL, MU, YP, KI, SE; all States except WA and NT.

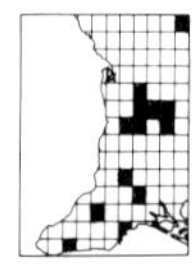

11. *Glossodia major*, **purple cockatoo**: 10-25 cm tall; fls rarely white; 'petals' 20-25 mm long; labellum (11a) 10-11 mm long. NL, SE; all States except WA and NT.

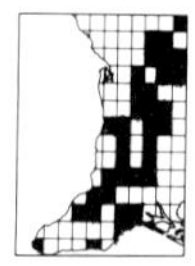

12. *Lyperanthus nigricans*, **red-beak orchid**: 10-30 cm tall; lvs very broad; 'petals' often c. 25 mm long; column (12a) c. 12 mm long. FR, EP, NL, MU, YP, KI, SE; all States except NT and Qld.

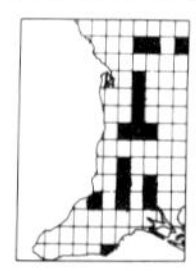

13. *Spiranthes sinensis*, **ladies tresses**: 15-45 cm tall; infl. fairly dense and with a characteristic spiral arrangement to the fls; fls (13a) white or pink; 'petals' 4.5-5.0 mm long. KI, SE; all States except WA and NT; New Zealand.

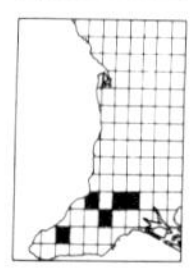

(*Continued*)

2
4
3a
5
7
6
1
3
7a
8
9
10
11
12
13
12a
8a
9a
10a
11a

Plate XCVI

ORCHIDACEAE
(Continued)

1. *Microtis parviflora*, **slender onion-orchid**: 7-50 cm tall; fls (1a) with a 2-5 mm long ovary; labellum (1b) 1-2 mm long. FR, EP, ?KI, SE; all States except NT; New Zealand.

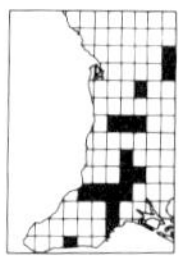

2. *Microtis unifolia*, **common onion-orchid**: 5-80 cm tall; fls (2a) with a 2-10 mm long ovary; labellum (2b) 1.2-2.5 mm long. FR, EP, NL, MU, YP, KI, SE; all States except NT; western Pacific. 3 other species.

3. *Orthoceras strictum*, **horned orchid**: 15-45 cm tall; the 2 long narrow 'petals' c. 2.7 cm long. FR, EP, NL, YP, KI, SE; NSW, Vic., Tas.; western Pacific.

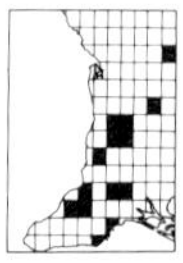

4. *Thelymitra antennifera*, **rabbit-ears**: 12-20 cm tall; 'petals' 15-16 mm long; column (4a) 5-10 mm long. FR, EP, NL, MU, YP, KI, SE; WA, Vic., Tas.

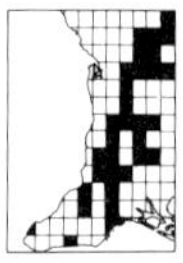

5. *Thelymitra aristata*, **giant sun-orchid**: 35-95 cm tall; 'petals' 13-20 mm long; column (5a) 6-7 mm long. FR, NL, SE; NSW, Vic., Tas.

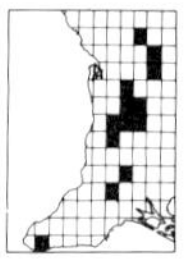

6. *Thelymitra flexuosa*, **twisted sun-orchid**: 15-20 cm tall; 'petals' c. 7 mm long; column (6a) c. 4.5 mm long. EP, KI, SE; all States except NT and Qld.

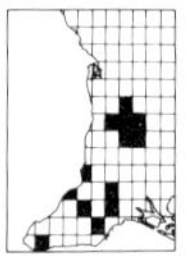

7. *Thelymitra benthamiana*, **leopard sun-orchid**: 15-45 cm tall; 'petals' 15-16 mm long; column (7a) 7-8 mm long. EP, KI, SE; WA, Vic.

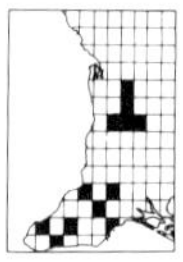

8. *Thelymitra ixioides*, **spotted sun-orchid**: 20-60 cm tall; 'petals' 8-9 mm long; column (8a) c. 4 mm long. SE; all States except WA and NT.

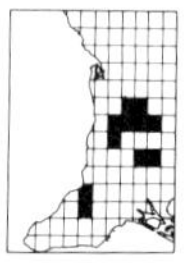

9. *Thelymitra luteocilium*, **early pink sun-orchid**: 15-37 cm tall; 'petals' 7-8 mm long; column (9a) 4-5 mm long. FR, EP, MU, YP, KI, SE; NSW, Vic., Tas.

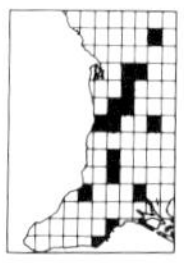

10. *Thelymitra nuda*, **scented sun-orchid**: 10-60 cm tall; fls often scented; 'petals' 10-20 mm long; column (10a) c. 6 mm long. FR, EP, NL, MU, YP, KI, SE; all States except NT.

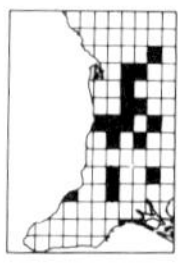

11. *Thelymitra pauciflora*, **slender sun-orchid**: 5-50 cm tall; 'petals' pale blue, lilac-blue or white, 7-10 mm long; column (11a) c. 5 mm long. FR, EP, NL, MU, KI, SE; all States except NT; New Zealand.

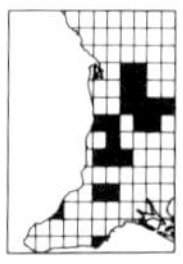

12. *Thelymitra rubra*, **common pink sun-orchid**: 15-50 cm tall; 'petals' salmon, pale pink or rarely pale yellow, 7-12 mm long; column (12a) c. 5 mm long. FR, EP, NL, SE; NSW, Vic., Tas. 8 other species.

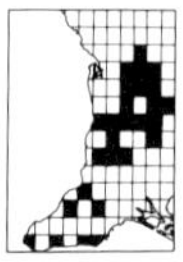

1
1a
2
2a
3
4
4a
5
5a
6
6a
7
7a
8
8a
9
9a
10
10a
11
11a
12
12a

Gazetteer

The grid reference refers to the map on page 8. These localities were extracted from specimen labels used in plotting the distribution maps. Spellings have not been updated.

ADELAIDE	F8
ALBERTON	E6
ALDGATE	G9
ALDINGA BAY	K5
ALDINGA Beach; Township; Scrub	J6
ALTONA SCRUB	B11
AMBLESIDE	G10
ANGLE VALE	B8
ANGOVE'S SCRUB	D9/E9
ANSTEY HILL Regional Park	E9
ASHBOURNE	J10
ASHTON	F9
ATHELSTONE	E9
AULDANA	E9
BACK VALLEY	M7
BAKER'S GULLY	H8
BALD HILL(S)	M5
BALHANNAH	F10
BAROSSA RANGES	A12/B12
BEAUMONT, including Common	F8
BEEFACRES	E9
BELAIR, including Recreation Park	G8
BELVIDERE	J12
BIGGS FLAT	G10
BIRDWOOD (Blumberg prior to 1918)	D12
BIRKENHEAD	D6
BLACK HILL Conservation Park	E9
BLACK SWAMP	L10
BLACKISTON, or BLAKISTON	G11
BLACKWOOD	G8
BLANCHE POINT	I6
BLATCHFORD'S SCRUB	M7
BLEWITT SPRINGS	I8
THE BLUFF (ENCOUNTER BAY)	N8
BOATHARBOUR CREEK	N4
BOLIVAR	D7
BONYTHON, Eric, Conservation Park	N4
BOTANIC PARK & Gardens	F8
BRADBURY	G9
BREAKNECK HILL	D10
BRIDGEWATER	G9
BRIGHTON	G7
BROWN HILL CREEK Recreation & Conservation Park	F8
BUCKLAND(S) PARK	B7
BUGLE RANGES	H11
BULL CREEK	I10
CAMDEN	F7
CALLAWONGA	N5
CAPE JERVIS	N2
CAREY GULLY	F10
CARRICKALINGA	K4
CENTENNIAL PARK	F7
CHAIN OF PONDS	D11
CHANDLER'S HILL	H8
CHARLESTON incl. Conservation Park	F12
CHERRY GARDENS	G8
CHERRYVILLE	E10
CHITON ROCKS	M8
CLAPHAM	F8
CLARENDON	8
CLAYTON	L11
CLEARVIEW	E8
CLELAND RESERVE or Conservation Park	F9
COBBLER'S CREEK	D9
COROMANDEL VALLEY	G8
COX'S CREEK	G9
COX'S SCRUB	J9
CRAFERS	F9
CROMER incl. Conservation Park	D12
CROMPTON'S SCRUB	M6
CUDLEE CREEK	E10
CURRENCY CREEK Township	L10
DASHWOOD GULLY	H9
DEEP CREEK	N2
DELAMERE	M2
DERNANCOURT	E9
DEVIL'S ELBOW	F9
DINGABLEDINGA	J8
DOVER GARDENS	G7
DOUGLAS SCRUB	I7
EAGLE ON THE HILL	F9
ECHUNGA	H10
EDEN HILLS	G8
EDINBURGH SWAMP	K7
ENCOUNTER BAY	M10
ENFIELD incl. Enfield Scrub	E8
ERIC BONYTHON Conservation Park	N4

ETHELTON	D6
EVANSTON	B9
FAIRVIEW PARK	D9
FERGUSON Recreation Park	F9
FILSELL HILL	F10
FINNISS, Township; Railway Reserve	K10
FLAXLEY	H10
FOLLAND PARK	E8
FORESTON	D11
FOREST RANGE incl. property of Field Naturalists' Society	F10
FULHAM	F7
GANDY'S GULLY	F9
GAWLER Cemetary	A9
GAWLER Township	B9
GILLMAN	E7
GLANVILLE (or Fort Glanville)	E6
GLENELG	F7
GLEN OSMOND	F8
GLEN SHERA SWAMP	K7
GLYNDE	E8
GOLDEN GROVE	D9
GOLDING'S SWAMP	N5
GOOLWA Township	L10
GOOLWA Scrub	M10
GRANGE or The GRANGE	E6
GRANGE Golf course	E7
GRANITE ISLAND	M8
GRASSTREE HILL	N4
GREENACRES	E8
GRÜNTHAL (since 1918, VERDUN)	G10
GUMERACHA	D11
HACKHAM	H7
HACK'S BRIDGE	H10
HAHNDORF	G10
HALE Conservation Park	C11
HALLETT COVE Conservation Park	G6
HALL'S CREEK	M7
HAPPY VALLEY	G7
HARDY'S SCRUB	I7
HARROGATE	F12
HARTLEY	J12
HEATHFIELD	G9
HENLEY BEACH Dunes and Foreshore	F6
HENLEY BEACH Suburban	F7
HERMITAGE, Upper and Lower	D10
HEYWOOD PARK	F8
HIGHBURY	E9
HIGHERCOMBE	10
HINDMARSH FALLS/TIERS	L7
HINDMARSH VALLEY Township	L8
HOPE VALLEY	E9
HORSNELL'S GULLY	F9
HOUGHTON	D10
HUMBUG SCRUB (Now PARA-WIRRA)	E10
INGLE FARM	E8
INGLEWOOD	D10
INMAN HILL	L7
INMAN VALLEY Township	L6
IRONBANK	G9
ISLINGTON	E7
JUPITER CREEK	H10
KAISERSTUHL (/ MT KITCHENER)	A12
KANGARILLA	H8
KANGAROO CREEK Dam/Reservoir	E10
KENTON VALLEY	E11
KERSBROOK Forest	C11
KERSBROOK Township	D11
KESWICK	F7
KING'S POINT nr KING BEACH	N8
KINGSTON PARK	G7
KLEMZIG	E8
KNOTT HILL	I9
KOOYONGA	F7
KUITPO Forest & Township	I9
KUITPO Colony	J9
KYEEMA Conservation Park	J9
LANGHORNE'S CREEK	J12
LARGS BAY	D6
LEFEVRE PENINSULA	D6
LENSWOOD Agricultural Research	F11
LINTON (or LYNTON)	G7
LIRABENDA Reserve	G10
LITTLE GORGE	L4
LOBETHAL	E11
LOFTIA PARK	G9
LONG GULLY	G9
LONGWOOD	G9
LOWER HERMITAGE	D10
LYNDOCH	B11
MACCLESFIELD	I11
MAGILL	E9
MANNING RESERVE	I8
MARBLE HILL	F10
MARDEN	E8
MARINO; MARINO ROCKS	G7
McHARG'S HILL	I9
McLAREN FLAT	I8
McLAREN VALE	I7
MEADOWS	I10
MEASDAY'S HILL	F9
MENGLER'S HILL	B11
MERRYBANK	L9
MEYER'S SCRUB	J10
MIDDLE BEACH	B5
MILANG	K12
MILE END	F7
MILLBROOK, Reservoir Catchment	D10
MOANA Beach/Township	I6
MODBURY	D9
MONIER SANDS, (or WHITE SANDS) Private Quarry	B12

Index

DYMOCKS Standard Bk
07/10/94 09:53 P 0 9493
1 @ 39.95 0864173237 $ 39.95*
PLANTS OF THE AD

SUBTOTAL	$	39.95
SALES TAX @ 0.00%	$	0.00
TOTAL	$	39.95
TENDER B/CDM/C	$	39.95

Thank you for shopping at Dymocks
Please call again.